Successful Business Operations for Electrical Contractors

Ralph Edgar Johnson

Engineering Graduate, Colorado School of Mines
Member of NECA Academy of Electrical Contractors
Author of Electrical Construction Cost Manual *and* Electrical Contracting Business Handbook
Past President
Sturgeon Electric Company of Denver, Colorado

Gene Whitson

Certified Public Accountant
International Accountants Society
Past Vice President
Cannon Wendt Electric Company of Phoenix, Arizona

McGraw-Hill, Inc.

New York St. Louis San Francisco Auckland Bogotá
Caracas Lisbon London Madrid Mexico Milan
Montreal New Delhi Paris San Juan São Paulo
Singapore Sydney Tokyo Toronto

Library of Congress Cataloging-in-Publication Data

Johnson, Ralph Edgar.
Successful business operations for electrical contractors / Ralph Edgar Johnson, Gene Whitson.
p. cm.
Includes index.
ISBN 0-07-032655-X
1. Electric contracting—Management. I. Whitson, Gene.
TK441.J645 1993
621.319'24'068—dc20 92-18194
CIP

1 2 3 4 5 6 7 8 9 0 DOC/DOC 9 8 7 6 5 4 3 2

ISBN 0-07-032655-X

The sponsoring editor for this book was Harold B. Crawford, the editing supervisor was Caroline Levine, and the production supervisor was Pamela A. Pelton. This book was set in Century Schoolbook by McGraw-Hill's Professional Book Group composition unit.

Printed and bound by R. R. Donnelley & Sons Company.

This book is designed to present accurate and authoritative information and to reflect the state of the law and regulations as they existed at the time of publication. Major changes may result from the future actions of code makers, labor legislators, Congress, state legislators, or the courts.

In recent years, the role of women in construction and engineering has seen tremendous growth, and this is indeed desirable. However, the terminology in these fields is replete with such words as *foreman, draftsman, workman,* and *craftsman*. In addition, the English language has no neuter personal pronoun. Therefore, where words such as *foreman* or the pronoun *he* are used, it is hoped that readers will understand them to refer to persons of either gender.

This book is dedicated to the electrical contracting industry, for whom the text was prepared and for whom it is hoped a great deal of benefit will accrue.

The cooperation and understanding of our dear wives, E. J. and Carolyn, were greatly appreciated.

Contents

Preface

One of the most important qualities required of an electrical contractor is confidence—the kind that is involved when a bid is submitted on a job that he feels certain will be successful and profitable and within the financial capabilities of his own business. This kind of confidence can result only from a considerable knowledge of the business and good judgment in its application.

The authors of this book have been on the firing line of the electrical contracting business for some time. Each has been a manager of a sizable business and as such has put his signature on seven-digit bids for difficult jobs, and has seen those jobs through to fruition. Each of the authors has prepared manuscript for sections of this book for which he was particularly well qualified, and as a result the information offered here should be practical, realistic, and useful.

It would be unfortunate in the highly competitive business of electrical contracting not to use all the modern, high-tech systems and equipment which aid management today in the operation of the business, but their full value cannot be realized without first developing a complete business system.

The most important of these high-tech business tools is the computer, and this book outlines how this device may be utilized and what is involved in its application.

Scheduling material for an electrical construction job, for example, is a very important part of the contractor's responsibility, and it involves preparation of the estimate in a manner which is useful for that purpose and in a form suitable for immediate use. The computer is extremely important for the purpose of furnishing the information with which the planner can work when he is making up the schedule.

It should be recognized at the outset, however, that all persons are not necessarily system-oriented any more than all persons are mechanically inclined—but that these skills can be developed in most cases, and that the most essential ingredients are knowledge, attitude, and desire.

The computer makes it possible through appropriate programs to utilize a complete business system in which all the details of the business are interrelated, and a great deal of the otherwise frustrating detail is handled with comparative simplicity and considerably more speed than by a manual system.

All the detailed information required for the development of a complete computerized business system is presented in this book in a manner useful for the successful operation of the kind of business which will be able to survive in the twenty-first century. Although opportunities will be abundant, the level of competition will enforce excellence in performance. The fact is that operating in this manner will provide peace of mind and a type of enjoyment in the workplace that is most reassuring.

In order to best utilize new technology, however, the manual system must be developed first and then applied to the computer. Important principles of business management, financial management, and marketing are also presented in this book in a thorough but uncomplicated manner. Controls and appropriate report forms are emphasized as an important factor in successful operations for electrical contractors.

Considerable emphasis is placed on the necessity of securing a required amount of sales which include an adequate amount of margin to cover overhead and profit. We also present a complete concept of how to manage the funds required to support the proper amount of business.

Successful operations don't just happen; they depend on the management of a myriad of details. But with a good knowledge of the business, and a properly planned operation, there is every reason to expect success. The future has never been brighter for those who prepare for it.

Dun & Bradstreet proclaim that the majority of business failures in our country are the result of two factors, poor management and inadequate finances. One of the absolutely essential functions of management is to make sure that there is a large enough volume of sales to sustain a profit. There is no question that there are many qualified electrical technicians in the business as contractors who are capable of solving any electrical problem presented to them—but while this type of capability is required of a contractor, the most important ingredient is to be able to operate profitably. Marketing is essential to profitability but it is also essential to know how to be a competitive bidder and operate successfully. Controlling a few job management factors enables the contractor to provide field backup and boost productivity.

We are indebted to the Estimatic Corp. of Denver, Colorado, for many illustrations, and to the National Electrical Contractors Association for many statistics and other important information.

Ralph Edgar Johnson
Gene Whitson

Introduction: The Key to Meeting Competition

Of utmost importance to the American system of free enterprise is competition. Competition provides lower prices, better products, better services, and higher productivity, which result in higher living standards. Competition provides a constant incentive to improve business operations and instills a continuing consciousness in both management and labor of the necessity to excel in their work.

The key to meeting competition is to develop an organization which is capable of buying material at the best prices, of showing superior ingenuity and enterprise in making installations at the lowest labor cost, and or providing maximum support for the men in the field to work within the limits of minimum cost operations.

The process of developing such an organization is not an easy one, but it must be done in order to operate successfully. The adage "Build an organization and the organization will build the business" is undoubtedly true provided there is proper leadership—leadership which establishes a complete operating business system, in which the organization is thoroughly trained, and from which are established the goals toward which to work.

Certain types of competition are not desirable to meet. Good operations dictate that a contractor must set his own standards; he must know his own limits and capabilities as to how far he can go to be competitive, and when the limits are not acceptable he will willingly withdraw.

Fortunately, knowledgeable operators usually set the pace because their limits are within their own grasp, and exceed those of less capable competitors. It is the purpose of this book to define the *intelligent* limits of competition and how to achieve them.

One of the principal factors in meeting competition is being able to obtain a high degree of productivity from the work force. In the business of electrical contracting maximum productivity can be achieved when the different specialists within the organization pool their collective efforts in order to coordinate estimating, job management, and purchasing with the installation in the field.

It is the contractor's responsibility to provide maximum field support. No matter how conscientious the field man might be, there are many things over which he has little or no control, but which lie in the contractor's control. The fact that labor is the foremost ingredient in installation work is undeniable—but if productive man-hours are to be achieved, it is mandatory for the contractor to control the flow to the job of the right type of material, in the right quantity, at the right time.

For a business to achieve outstanding levels of performance, it is necessary for it to be staffed by individuals who have been properly trained to perform the task to which they have been assigned, and for whom suitable forms of communications are provided so that one specialist can pick up the work where another leaves off. One useful dictionary definition of *productivity* is "the capability of effecting or contributing to effective production." There is a tendency in our industry to think of productivity as a high degree of production in the field. While this is of predominant concern to the contractor, it is not the only important aspect of productivity. The U.S. Army Corps of Engineers defines *productivity* as "the capability to perform a given task in a specified time, at a reasonable cost, and with a minimum of waste." A reasonable cost is established by competitive bidding in the marketplace, and the elimination of waste is achieved by the efficient operation of the contractor and is considered when determining the cost—so, in a sense, the real measure of productivity is the capability of the contractor's organization, in concert with the field personnel, to compete in the marketplace and successfully meet the cost requirements of the job.

In industry, *productivity* is defined as "the ratio of effective work actually done to that which could have been done under standard conditions which are uniform and well established by usage." This definition emphasizes that productivity can be greatly improved by better production methods. In the electrical contracting business we can effect this type of improved productivity by employing such practices as preassembly and prefabrication.

The limits imposed by the nature of electrical construction work—by the fact that it is necessary to fit our systems into a structure by hand, piece by piece, over large areas of space—greatly inhibit major improvements in productivity; therefore it is necessary to continually attempt to make those improvements in productivity which *are* feasible.

The definitions of productivity which we have considered emphasize the requirement of capability on the part of the contractor. *Capability* relates to overall efficiency, joined to physical and mental ability. It should be recognized that a contractor or journeyman could have physical and mental ability without being efficient, and it should be apparent that an industry heavily populated by this type of person would be hard pressed to achieve a very high degree of productivity.

Efficiency is defined as "effective operation as measured by a comparison of production with cost." It appears that this is the crux of the whole matter of productivity—to be able to bring about effective production while keeping costs competitive.

Probably the most important aspects of the contractor's operation in terms of meeting the requirements of effective production are planning and scheduling. In order to plan and schedule effectively, there must be available a plethora of information about all phases of the job at hand. The information will be provided by an accurate, detailed estimate which is broken down into sections of the job, which correspond to natural subdivisions of electrical construction work. All the scheduling done for electrical construction work must be predicated on the general contractor's construction schedule for the job, and the accuracy of the electrical schedule will be largely governed by the accuracy of the general contractor's schedule. An accurate schedule makes it possible to assign requirement dates to material purchase orders with confidence that the material will be on the job when it is needed, and also that manpower and tool requirements can be anticipated with accuracy.

Planning is the responsibility of job management, which also provides field support for the job, one important aspect of which is selection of good supervision and workmen. Job management also prepares the plans for field installation, develops an appropriate Bill of Material, and provides on-the-job facilities to accommodate material, tools, and equipment handling and storage.

The ability to meet competition is greatly influenced by the confidence of an electrical contractor in the bid price assigned to the job, and this confidence is the result of knowing that the organization is capable of good productivity on the job after the bid is successful and the job is under way.

Because the amount of important detail involved in the business of electrical contracting is so vast, it is essential to utilize the computer in all operations, including planning, organizing, activating, and controlling. The computer is the most important tool available to management for managing time, and making every employee's time count is mandatory if success is to be achieved by meeting competition. The use of computers in the business ensures proper handling of detail in the process of estimating, automatic production of a Bill of Material designed for scheduling and purchasing, easy comparison of actual to estimated job costs, timely reports on profit and loss and many other accounting functions, and a vast amount of operating control information, which would, without the computer, be available only to a large, sophisticated business operation.

In order to utilize computers properly, it is essential that the individuals involved be system oriented because each action involves the handling of detail. Unless a person has a natural aptitude for handling detail, it is necessary to train them to do so, and this can be done when they are indoctrinated in the system.

Meeting competition is within the grasp of all electrical contractors who are system oriented and utilize all the management concepts available today—including those presented in this book. Management must have expertise in the electrical functions of the business in order to be effective, but it is even more crucial that management have expertise in the operations which make it possible to be successful.

Chapter

1

Effective Business Management

The primary requirements for successful management may be compared to the abilities and responsibilities required of an airline pilot. The pilot must plan the route to allow him to arrive at his destination, and the manager must make plans which result in the achievement of company goals. In both instances the planner must determine a sequence of events necessary to accomplish the goal. The pilot must have a set policy and set procedures, such as the strict requirement to perform his checklists and preflight procedures to assure the safety of the flight. The manager must include, as part of planning, both setting the overall objectives, and setting the policies and procedures to establish acceptable standards of performance.

The pilot and manager must each organize to determine the functions necessary and in what order they must be carried out in order to arrive at the destination. The authority of and responsibilities of each member of the crew under him must be defined. The manager also is responsible for selecting and assigning personnel, and providing for indoctrination, training, and retraining.

Both the manager and the pilot must practice the art of leadership to motivate personnel and guide their thoughts and actions in a responsible and intelligent manner. *Activating* is the process of motivating personnel to achieve company goals; it is the art of guiding the thoughts and actions of others in a responsible and enlightened manner. The leader must direct personnel activities so that they will be carried out in a cooperative and harmonious manner.

The pilot and the manager must be in constant control of the plane or the business. The pilot observes the weather and the performance of the plane and trims it for the most efficient and comfortable operation.

He monitors his instruments to assure himself that the plane and all its various systems are functioning correctly and that he remains on the planned course. The manager constantly observes the overall operation of the business, paying particular attention to any complaints from customers, suppliers, or employees, and makes any adjustments necessary to improve business performance. The manager's instruments are the reports on the business operation. These reports enable him to measure actual performance against planned performance and take corrective action as required.

A successful business which provides outstanding services is always staffed by individuals who are able to perform in unison the many tasks necessary for the attainment of company goals.

Planning the Business

The planning for a successful electrical contracting business must start long before the doors are open and the first contract is obtained. The person or persons contemplating starting such a business should ask themselves some questions to determine their personal suitability for starting the business and successfully managing it.

Many electrical contracting shops are started by journeyman electricians who have little business experience and have spent their careers working for someone else. If this describes you, you should ask yourself the following questions:

1. Do you respond well to authority?
2. Do you like routine work patterns?
3. Do you enjoy a stable, constant work situation?
4. Do you work well within clearly set parameters?
5. Do you work hard to be in a "comfort zone"?

If the answers are yes or nearly yes to most of these questions, you probably will have problems running your own small business because of the uncertainty of the work environment.

Another set of questions which directly relate to the personality traits that impact on the ability to run a business are:

1. Do you look at many problems as opportunities in disguise?
2. Do you think that work is an all-important thing in your life?
3. Do you need to be in control?
4. Do you seek change in variable situations?

5. Are you willing to take quick action when called for without getting approval from other people?
6. Do you enjoy establishing the parameters and objectives which you work by?
7. Are you constantly thinking about new things to do and avoiding routine?

If your answers to these are positive, it is a good indication that you are a promising candidate and probably suited to own and run your own small business.

The next step in this preliminary planning process is to determine what kind of business you are best suited for. Ask yourself:

1. In what kind of business or work do you have the most experience?
2. During the course of your life what is it that you have really done well?
3. What have you done poorly?
4. What is it that you love to do? Work with your hands, plan a job, or direct others to do the job?
5. What kind of work do you love the most? Will this business provide this for you?
6. Is there a market for your kind of business in the area where you propose to operate?

There are several different approaches that you can take when starting your own small business. An electrical contracting business which is run out of the owner's home is rarely successful. Though a home office operation does not have to pay rent for an office and yard, local zoning laws may prohibit running a business in a residential neighborhood, and the neighbors are not likely to appreciate having an electrical contractor's yard and equipment near their residences. The investment required for a business located in the home is small, and this reduces the risk of large losses, but there is little chance for real growth. The owner of the business in most cases will find that they are working for regular wages or less, and that they have acquired many additional responsibilities such as payroll taxes, business license, bookkeeping and accounting, equipment maintenance, and marketing and sales that they did not have while working for someone else.

The business can be started by investing your own funds or borrowed funds and renting or buying a location suitable for the business. This approach requires the hiring of employees and acquisition of equipment. There will be a certain lead time before the business produces a profit, but if the business has been well planned, has astute management, and has sufficient capital, there is a good possibility of capital gain.

Another possible way to enter the electrical contracting business is to buy an existing business. Considerable capital is generally required, but with proper study of the past record of the business and of future market prospects the risk is moderate, availability of borrowed capital is easier, and growth with income is probable.

One of the main obstacles to starting a new small business is obtaining adequate financing. Different types of small businesses require different amounts of start-up capital. An electrical contractor doing only electrical service would require less capital than a contractor involved in medium to large construction contracts.

There are several ways of funding a new business, ranging from using your savings to borrowing most of the capital from a lending institution. Some of the financial options open to you are:

1. *Equity money:* Funds out of your own pocket. Some of the equity money may be borrowed. Funds would be available for borrowing by putting a mortgage on your home. A second mortgage or home equity loan carries relatively low interest, and in many cases most of the start-up money will have to come from this source.
2. *Outside investors or sources:* If you do not own a home or have the needed amount of available cash, you may find that friends or family may be willing to invest in your business.
3. *Commercial banks:* A commercial bank may be willing to fund your business. There are consumer finance organizations that are able to supply either short- or long-term loans in smaller amounts. In most cases the consumer finance companies will demand a higher interest rate than regular banks.
4. *Savings banks and savings and loan organizations:* Like commercial banks, savings banks and savings and loans will sometimes make loans for business start-up.
5. *Government agencies such as the Small Business Administration*: The Small Business Administration (SBA) guarantees the loans to a bank or other lending institution. The paperwork involved in qualifying for an SBA loan is daunting to the average person and the agency has stringent requirements which must be met before a loan will be granted. The government guarantee can mean that you can get a lower interest rate and larger amount of money than you could get without it.

At this point you must decide if owning your own business is what you really want and are equipped to do. Review the three sets of questions at the beginning of the section to arrive at a conclusion.

Once you have decided you do want to start a business, the next step is to formulate your plans and write them down in the form of a business plan and a marketing plan. Planning for success requires several steps:

1. Determine the kind of business you should be starting. Do only work that you love doing. Excellence and motivation usually result from doing satisfying and enjoyable work. Make sure there is a real market for the product or service you plan to offer. Does your available financing provide sufficient funds for this business?
2. What form should your business take? There are three forms: sole proprietorship, partnership, and corporation. Do you want to do this alone as a sole owner and proprietor with all the responsibility? Is there someone who would make a compatible and responsible partner, with the same interests and goals as yours? A corporation has advantages and disadvantages. Both the corporate form of business and the partnership can provide medical plans, profit-sharing plans, and other legal tax breaks. However, the corporate form can result in double taxation unless careful tax planning is in place. Using the corporate form, if properly structured, can prevent your having personal liability for corporate debts. To form a legal partnership or corporation the services of an attorney should be used.

The formulation of a sound business plan can greatly aid in the success of an enterprise. The lack of good planning will almost surely lead to failure.

It is during the planning stage that a firm's objectives and direction are established, and these objectives must be based on hard facts and not fantasy. When determining the objectives of the business, it is necessary to make sure there is a need and market for your services or product. What is a reasonable time to expect to wait before the business turns a profit? Many studies over the years have proven that in most types of new business at least a year of operation is required before the business will produce a profit.

Of primary importance is a study of the available market. What area do you plan to serve? Is the market in that area a growing or declining one? Is your planned market area large enough to provide the business volume you need in order to make a profit?

Prepare everything in the business plan in writing. The following information should be included:

1. Start out with a description of the business you are going to start and a statement of the purpose of the business.
2. Set up a formal table of contents.
3. Provide a detailed outline of the products or services you are going to provide.
4. Describe the location of the business and if possible the available opportunities for growth.
5. Define the management of the business: Who is responsible for the success of the business.
6. List personnel who are going to do the work and sources of additional personnel.
7. Create a marketing plan. What market do you plan to serve and what are the demographics of that market? What are the preferences of your market and how do you propose to meet these preferences?
8. What will be your competition? Who else is in the same business trying to get market share.
9. What will be the price structure of the business? Will the business be serving the upper, middle, or lower end of the market?
10. Make a statement concerning quality: How do you propose to control it? The quality side of the market is growing most rapidly.
11. Consider both the short-term and long-term trends in the market.
12. List what share of the defined market you expect you will be able to get and what projections you have made to attain these goals.
13. Describe in detail the financial arrangements of the business. Begin with an account of the sources and applications of funding: Where is the money going to come from? Are you going to provide some yourself, borrow some from banks, outside investors, or loans from family and friends? How is the money going to be used? For inventory, equipment, working capital, or toward the purchase of real estate?
14. Prepare a capital equipment and capital improvement list. What kind of equipment are you going to buy and exactly what is that equipment going to do?
15. Prepare an opening balance sheet which lists all the assets and liabilities of the business.

16. Prepare a break-even analysis showing how much business you have to do before you have profits.
17. Include income projections showing at least 3 years, by the month, and noting your expected sales, profits, and projected expenses.
18. Someone in the organization needs to know how to read a financial report and interpret the statements. List this person or firm.
19. Do not work with the tools.

Management's Duties and Responsibilities

Any business organization must produce acceptable goods or services, and must have the ability to market them at a profit. These requirements are basic to the success of a business, and management must accept the following responsibilities to meet them:

1. To develop products and markets that will provide the profits necessary for the survival of the business.
2. To make the organization's resources—human, material, and financial—work together in the most profitable way possible.
3. To train and retrain qualified people to enable them to step into management positions.

Accurate planning for making the most effective use of the company's human, material, and financial resources must include recognition of the economic potential of the area being served, budgeting for sales and other overhead costs, and assessing the company's ability to meet or exceed its objectives.

Effective use of the organization's human resources involves making provision for the succession of qualified people to perpetuate the business. It is the manager's duty to train at least one other person as a successor—and hopefully this person will be even better at the job than the present manager. Successful executives recognize the fact that their ultimate success depends on the skills and abilities of those under them. The manager who attempts to be all things to all people on the job and is fearful of thoroughly training his subordinates can hinder the future success of his company. For example, department heads require 5 years to properly prepare for their job, so they must be chosen long before the time they will assume their responsibilities.

The process of planning will result in policies and procedures that must be adhered to in order to meet the objectives of the company. To document these policies and procedures for constant reference, a manual of standard procedures should be established.

In the business of electrical contracting, the product is the installation of electrical work; the market includes

1. The construction industry
2. Commercial businesses
3. Industrial plants
4. Institutions
5. Utility industries
6. Specialty service industries
7. Public works

If the company is to earn a profit, management must develop a plan that will result in securing an adequate volume of business and a pricing policy which enables it to render adequate and competitive customer service, retire its obligations, provide properly for its employees, and accumulate assets for growth.

Pricing Policy

It is essential that a pricing policy that is competitive be established: this is necessary both to ensure good customer relations and to maintain management and staff's focus on the need for profit. A pricing policy must be adopted which will establish the proper reputation in the community. Electrical contractors must be thought of as being fair, honest, competitive, and successful in the conduct of their affairs.

In all matters of pricing, the objective is to sell at the lowest price compatible with making a sufficient profit to sustain the business and attain the company's goals. It is important to note that the nature of the electrical contracting business is such that as the company enters into larger dollar contracts, and the dollar volume of work performed increases, the obtainable profit margin as based on sales will be decreased. The National Electrical Contractors Association (NECA) publishes statistics on their members' operations every 2 years; for the year 1990 the summary of these statistics indicates profit margins for electrical contracting businesses of various sizes, based on sales volume, as follows:

Total sales	Profit as percent of sales
$ 711,800	4.58
1,476,811	3.76
2,830,161	3.75
6,042,106	3.49

The profit percentage depends on the types of jobs, the location of the business, material and labor costs, any special requirements for the job, and generally the ability of the individual contractors to perform efficiently.

The $20,000 job will require the contractor to use different labor rates than he uses on the $100 service call. First, service-type work requires travel time. Second, much service work involves remodeling and therefore requires removal and replacement of earlier work, as well as additional time used to select materials. For these reasons the contractor must increase his labor billing rate for that type of work. In some instances this rate will be double or more his total labor cost.

Material billings on small service-type work will also need to be higher than those on the larger work. This is because materials are purchased in smaller quantities for the small job, and therefore, the cost is higher. In some instances the contractor may have the material in his warehouse stock; if so, he has hidden costs in the form of storage and handling charges. These costs must be considered in arriving at a selling price.

In many cases it is much easier to bill service-type work at a high labor rate and a list material rate, which together are high enough to cover all the various hidden costs in addition to the required profit margin, than it is to convince the customer that you are entitled to a profit on your work.

On the $20,000 job the contractor will probably have competition in the form of a bid from another contractor, he should then use a labor rate which allows for some lost time, but not nearly as much as on the small service job. The gross profit on the $20,000 job will be an addition to the sum of estimated costs for labor and material. The gross profit add-on must be large enough to cover all overhead costs and a reasonable profit margin.

As the contractor grows and enters into large, multimillion-dollar contracts, the demands for accurate estimating and creativity in arriving at a proper price become acute. Competition will be more of a problem, and many times it will appear that the competitor is pricing below your cost—and he may well be.

In no instance is it advisable for you to lower your calculated prices just to meet competition. In cases such as this you must use ingenuity to arrive at the proper price which will be acceptable to the customer. The specifications must control, but design can be modified, good substitutions can be made, different material applications can be chosen, and technical and cost-effective engineering changes can be made.

Change Orders are a fact of life on larger contracts, and some contractors will enter a low bid on the base contract with the intention of marking up the Change Orders to cover some costs and most of the profit. This is not an acceptable practice in the eyes of the customer and the customer may discover the overcharges: then the contractor will suffer much embarrassment and probably lose the customer for any future work. In many instances, the base contract work has either been started or is finished before the Change Order is issued. If so, this fact must be recognized and incorporated into the pricing of the Change Order due to the "tear-out-and-replace" factor. The pricing of Change Orders should include margins and adders which are acceptable to the customer. These should be approved in writing by the customer before the Change-Order work begins.

On all Change-Order work, accurate cost records should be kept, the approved job factors applied properly, and only agreed-upon margins added. The almost standard margins for overhead and profit on Change Orders in the electrical contracting industry are 15 percent overhead and 10 percent profit. These margins present a loss problem for some contractors, and it is necessary in these instances to apply some markup to material and perhaps labor to fully recover cost and realize a reasonable profit.

In today's construction market many projects are negotiated before the project planning stage is completed. In these instances the electrical contractor is required to establish a budget price which is little more than an educated guess. This price must necessarily be high enough to protect the interests of the electrical contractor because it is very difficult to increase the budgeted price once the plans are published. It is advisable to attempt to stay within the customer's original budget figure providing that figure is reasonable. If the original budget figure is very low and the customer cannot or refuses to consider an increase, it is advisable to drop out of the negotiations.

When the budget figure is acceptable and you have been able to negotiate an agreement, it is still your responsibility not to betray the customer's trust: If cost savings are available they should be passed on to the customer. By doing so, you will enhance your reputation for honesty and earn the customer's confidence.

Organizing the Business

As the use of electrical energy and electrical equipment has grown over the past several decades, the electrical industry, in general, has grown and assumed a position of great importance in national commerce. The electrical contractor has been required to grow along with the industry; there has been a gradual transition from the one-man operation to a company with many employees and considerable equipment. The electrical contractor who has been successful during this transition has the ability to manage his operation efficiently.

The management organization of any company will become more complex in direct relation to the growth of the company. But regardless of the size of the company the policies controlling operating functions and the responsibilities of employees remain. The organizational structure will determine whether the responsibilities are vested in one person or in a management group. In the small management group each member will necessarily perform several functions, while in the large management group each member will have clearly defined individual duties.

The term *functional organization* refers to a company which has developed its management framework around the functions necessary to operate the business. In developing this management structure the company will have grouped related operations as parts of an important, defined business function. This system has the advantage of clearly defining the duties and responsibilities of each member of the management team, so that when one or more of them leave the firm, the basic management framework will remain intact and the specific duties of their successors will be clearly defined.

The functionally organized company is developed from the bottom up and the organization consists of the following major elements:

Operations. The front-line or production section of the business. Some examples are actual electrical installation, bookkeeping, material handling, and equipment maintenance. Each of these will be under the control of a supervisor.

Subfunctions. Certain operations such as accounting, material management, equipment management, and job management. The subfunctions may be performed by one person or a group having similar abilities, and in most cases will be under the direct control of middle management.

Major functions. The process of coordinating and directing the component operations as grouped by their relationships to one another and to the overall problem to be solved. Major functions would be handled by a company officer such as a vice president.

General management. The apex of the management framework, which coordinates the major functions and directs them in an effort to achieve the main business objective: profit. General management is the highest executive management level.

An example of how this type of organization works is as follows: The operation of bookkeeping involves keeping the general books of the company, maintaining payroll records, and preparing profit and loss statements. This operation, grouped together with similar or related operations, makes up the subfunction of accounting. The subfunction of audit is the process of checking all expense accounts, general records, and other documents for accuracy to assure that the accounting function has provided valid information to management. The information provided by accounting and verified by audit enables management to determine a business course, and thus the information becomes part of a major function called *control.* By coordinating this and other major functions from other phases of the business, management is then able to determine company policy.

An organization chart tailored to meet the specific requirements of a particular business is necessary to properly view the functional organization of that business. Viewing the business as illustrated on an organization chart and developing a comprehensive management chart will provide the following:

1. Lead to a recognition of all the component elements and their interrelation in a complete business operation.
2. Illustrate all the individual functions that are required for the profitable continuance and perhaps expansion of the business and show where additional personnel are required for expansion of the business.
3. Indicate the need to train personnel to replace a person now handling multiple functions should this person leave the organization.
4. Reveal possible problems which might be encountered as a result of expansion.
5. Uncover instances of overlapping authority, duplication of effort, unbalanced delegation of responsibility, and other improper conditions. Management can untangle such snarls by developing a chart, with functional elements properly segregated and correctly related, and by adding names of persons discharging specific duties in the proper areas. Here the chart serves as a corrective tool.
6. Give management a clearer picture of and closer control over the overall business operation.
7. Increase the overall efficiency of the operation.

Functional organization is based on logic and facts; it provides the only set of infallible rules for business progress and is a real asset to any business operation. A typical Functional Organization Chart is illustrated as Fig. 1.1.

The electrical contractor should review his business operation and develop such a management chart, whether it be simple or complex, to include all the functions and operations performed in his company. This will provide a complete picture of the organization.

A Standard Procedures Manual should be developed which establishes in detail all the functions and operations of the business, such

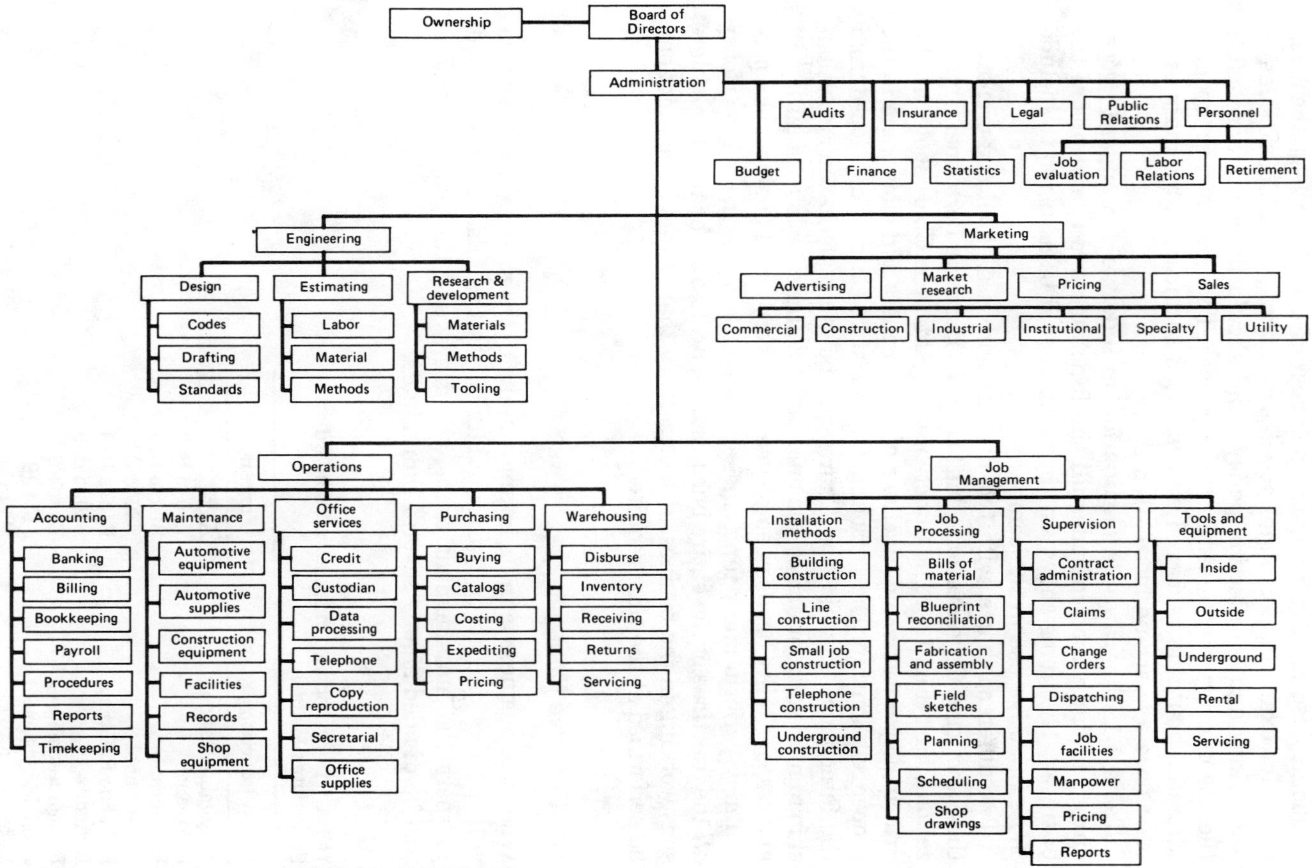

Figure 1.1 Functional Organization Chart for electrical contractors.

as the job description and duties and responsibilities of each employee. It should also include company policies on every phase of the business—for example, personnel policies on hiring, vacation, sick leave, and salary and a collection policy—and should illustrate and define the use of every form used in the business. The Standard Procedures Manual should be reviewed and updated to include any changes at least once a year and more often if required.

Table 1.1 illustrates an outline for an index to a Standard Procedures Manual which includes all the functions and operations of the business. The items shown in the table should be adjusted to include those in active use in the company.

Two different charts, with two distinct purposes, should be prepared: the Functional Organization Chart (Fig. 1.1) and the Personnel Organization Chart (Fig. 1.2). The Functional Organization Chart shows all the principal operations and functions of the business and plots them in the proper relationship. The Personnel Organization Chart is developed from the Functional Organization Chart and shows people in their relationship to management and to one another: It defines the lines of authority and responsibility between employees. In a very small business one person may perform many functions, and one or two persons may perform *all* the functions of the entire business. Regardless of the number of persons working in the business it is necessary that all the basic functions be performed at the time required.

The Effective Manager

When the business is in its infancy, many of the managerial functions will devolve on one or two individuals. As the business grows and prospers the management team will grow, and the responsibilities will be divided among several individuals.

TABLE 1.1 Example of Index for a Standard Procedures Manual

Function	Form no.
A–Administration	A1 to A9
B–Finance	B10 to B19
B–Insurance	B20 to B29
B–Personnel	B30 to B39
B–Public Relations	B40 to B49
B–Audits	B50 to B59
E–Engineering	E1 to E9
J–Job Management	J1 to J9
M–Marketing	M1 to M9
O–Operations	O1 to O9

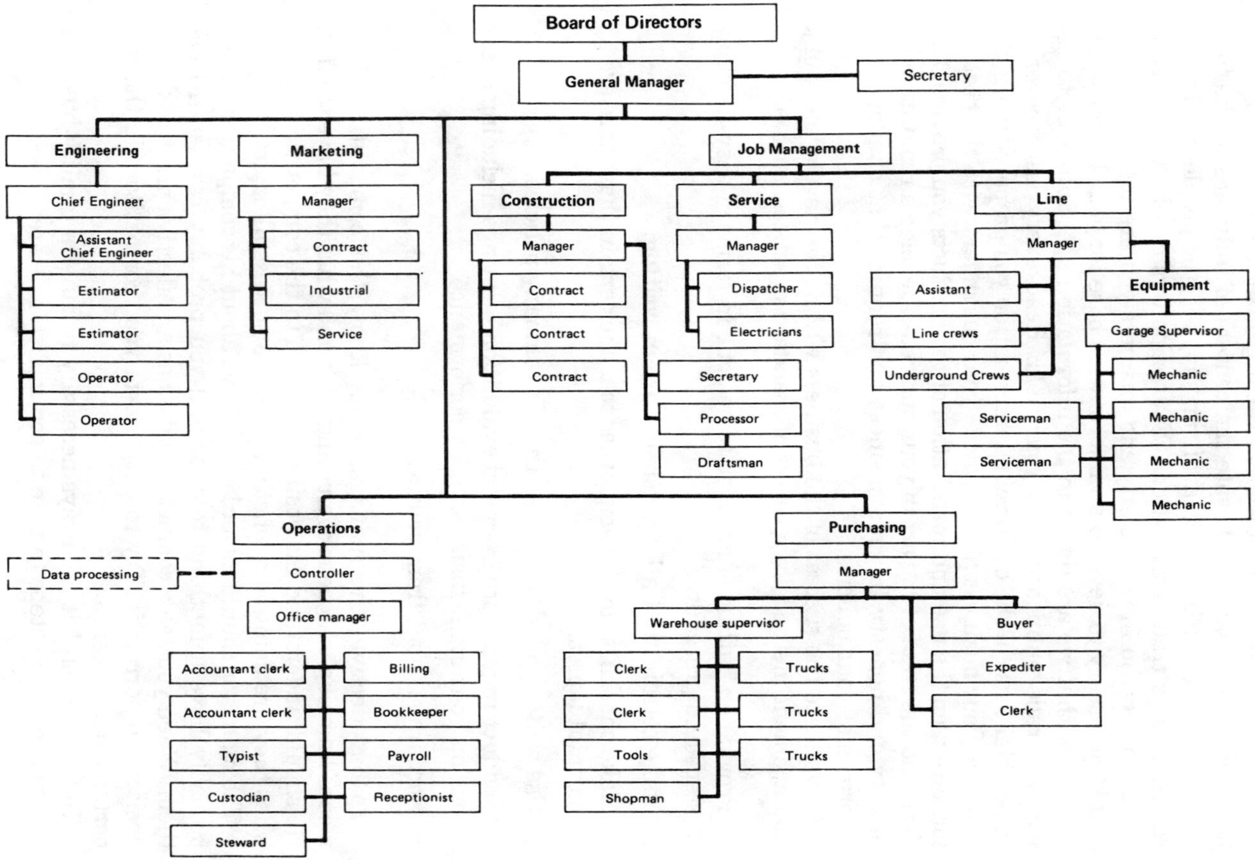

Figure 1.2 Personnel Organization Chart.

To be effective, the manager must be certain to have monthly budgets and operating statements prepared to allow him to be constantly aware of the financial condition of the organization.

The manager should cultivate the centers of influence in the field such as the accounting firm, law firm, supplier, the people involved in the industry's trade associations. Your suppliers can help you determine who is who in the locality and business you are in.

It is necessary to develop an information base for your business. Do not forget that we are living in the information age. Subscribe to the leading trade publications in your field. Read every issue cover to cover. Make use of the background information you prepared while developing your business plan. Provide your customers with the best information on the use and installation of the products you provide. Understand the state of the art in your business. Develop a reputation for quality and honesty in your business and require that employees be informed of your policies.

Of value to the success of the business will be the development of a management methodology, or set of standard practices, such as:

1. Provide an outline of each job to be done in terms of easy-to-follow steps and procedures.
2. Make certain that all agreements are in writing.
3. Make checklists for all employees to make certain that there are no harmful lapses.
4. Use written job descriptions that fit company policies.
5. Conduct regular progress and evaluation meetings including every one on your staff. Insist on good performance.
6. Encourage input from your staff.

To be effective, the manager must know the competition; that means knowing who he is competing with and what they have to offer. He should understand his position relative to the competition—know what they are doing and whether they are successful or not.

The manager should never be defensive about his competitors. They probably have something to offer and you probably can learn from them. When you see a competitor who has an effective practice do not hesitate to borrow it. Try to outperform your competitors in the important area of service.

To be successful it is always necessary to build a quality business reputation. Some steps to take to accomplish this are:

1. Retain only those employees who really care personally about the quality of the company and the products and services it provides.

2. Do not tolerate shoddy performance under any circumstances.
3. Design your products and services to meet the top of the market.
4. Stand behind your product and services. Never be satisfied with your performance.
5. Find and hire good employees and pay them well.
6. Improve your management skills constantly. You never know enough in a small business. Develop regular habits of study. Take courses every year to improve your skills.
7. Make yourself dispensable by developing strong employees capable of replacing you.
8. Test a new product before launching it. Be prepared to abandon losing programs and cut your loss.

The executive manager (sometimes referred to as *general manager*) must not only assume the role of leader of the business but also that of entrepreneur. He must determine which risks to take in the promotion of the business and will spend a considerable amount of his time meeting with those persons in positions of authority in other businesses who can provide opportunities for increased or new business. He should also spend a major portion of his time developing new contacts for future business opportunities. Therefore, the executive manager must have managers reporting to him who are capable of keeping the business running during his absence. For the functional organizational structure to perform effectively, it is necessary to keep managers out of the supervision level and the executive out of the middle management level.

The *supervisor* is charged with the correctness of the production work. He interfaces directly with the production workers who report to him. The supervisor must have detailed knowledge of the work of the production workers because he depends on them to train other workers without knowing if they are properly trained or not. He is confined to correcting mistakes after they happen rather than preventing the mistake in the first place. In order to be able to control the work, he must have detailed knowledge of the way the work should be done. He is the member of the management team who is paid to be up to date on the technology required for the job.

The executive manager, or general manager, should not be involved in the supervisor's job. He should not follow the chain of command downward to get an answer to a problem from the supervisor. If the executive manager follows the chain of command down to the supervisory level, there is a duplication of effort and this can become costly because the middle managers become merely messengers. In a well-run company, the intermediate manager will have only top supervision or other managers reporting to him.

The functionally organized business is a multilevel structure, with two different types of managers, the executive manager and the middle manager. Executive management has three components:

Entrepreneurial: Shall we expand, shall we buy that other company, should we issue stock to raise additional financing? External functions decide the success of the company. If the executive manager makes the correct decisions on markets and long-range planning, the company may survive even though they are not doing a good job.

Community relations—Public relations: Public relations does not mean serving on committees or charities. It is important to spend time tracking events in the local area which are going to impact the future of the company. Are there new development projects or public works in the planning stage to start in the future? It is important for the executive manager to constantly observe the conditions which are beginning to emerge which will affect the company.

Top-level liaison—Relationships with other executives: The third external function involves spending much time with other CEOs or government personnel who can control situations which affect the future of the company. Some top managers find themselves feeling guilty when they are out of the company spending time with other powerful people and not back at the store running things. If someone, and this has to be top management and not delegates, is not spending time on these external functions, sooner or later the organization is going to suffer.

From the above it is clear that the duties of top management are principally *external.* It is not the duty of top management to try to run the daily operation. The top manager should have middle management people reporting to him who are responsible for these functions and also coordinate with other same-level departments.

The job of middle management is *internal,* to run the work on a day-to-day basis. The middle manager should be responsible for resource or work management, manpower loading, scheduling material, and budgets. The middle manager has to understand the work going on about him but not the scheduling and examining of the workers' methods or hours. These functions are the responsibility of the supervisors. Examples of middle management personnel are the managers of the tool department, the material department, the estimating department, the accounting department, and the various production departments such as small contracts, industrial contracts, line contracts, and commercial contracts.

The second role of the middle manager is organizational management: running not only the work but also the organization. He should concern himself with the implementation of company policies handed down by executive management and with the processes the organization uses to perpetuate, rebuild, and adapt to current conditions. He should ascertain that communication lines are open and that communication flows both up and down as required.

The middle manager must determine that responsibilities are taken on and handled by the right people. He must also take steps to find out who the people are that are qualified and ready for advancement—and also to find out about the dead wood and prevent it from growing in the organization.

Middle management should make certain that the emergencies that are likely to occur are provided for in advance. The middle manager will be responsible for determining where the most effective manpower mix is in the organization, and whether there is a provision within the organization for continuous improvement in the work processes, communication process, and development process.

Middle management should concern itself not only with getting the daily work accomplished but also with making the organization better in the future. Middle management activities aimed at improving the organization fall into four major categories:

1. The first is cost reduction and profit improvement. The middle manager should make sure his supervisors are thinking about work simplification. He must take a look periodically at the methods being used by his supervisors in preparing their budget. Is the supervisor getting better in costing, setting up performance standards, and setting up his workload so that the projections he makes in the future are more reliable?

2. The second is maintaining and improving the organization's structure; the structure is constantly changing and it must be continuously reviewed, updated, and renewed. This involves systematically reviewing the job descriptions and mission description and comparing them to what is really going on in the organization. If what is really going on is different, then the manager needs to review the organization objectives and discuss them with subordinates, revising either the objectives or the official statements of responsibility depending on which needs to be changed. The middle manager should also periodically clarify the responsibilities of line and staff subordinates.

3. The third area is work force development. This involves identifying the qualified and high-talent people and blocking out the dead wood. The appropriate manager should inventory managerial and supervisory human resource reserves and personnel-building plans to assure that any spots in the organization that are going to need backup have several likely candidates.

Middle management needs to take a look at the performance appraisals done by the supervisors, make sure that these are of high quality, and then follow up on the relevant plan that arises out of them. If he wants employee development to take place in his organization, the middle manager must make it happen.

4. The fourth set of activities are those aimed at ensuring that each function of the organization is carried out by the person at the appropriate level of the organization. The middle manager needs to make sure that the foreman's job and subforemen's jobs get performed at the work site level and do not get diluted upward into the center of his organization, either to his own level or to that of the managers below him. He must test once in a while what is going on in his organization to make sure that the organization has not thrown up obstacles to the performance of supervision or that upper levels of management have not taken away supervisory authority in one way or another.

He should have his supervisors make an analysis of how they are using their time and why. What are the demands on their time and why do those pressures exist? Once he knows this, he can analyze how well the organization is going in allowing the supervisors to do the supervisory job. The middle manager should broaden supervisory authority and sense of responsibility by removing any staff approvals or control over supervisor's decisions. For example, he should not allow the personnel department to pass judgment on the supervisor's wage decisions, or the purchasing department to disallow the supervisor's signature authority, or the accounting department to require prior approval before the supervisor approves his workers' expense accounts.

In managing an organization we must recognize that the manager—the one who gets things done through several layers of supervision—has a unique function requiring work management and organizational management, and that it is at the middle of the organization that these functions should be performed—not at the bottom and not at the top. At the bottom the supervisor, not the managers, should be getting into the details of how the workers are doing their work. And at the executive level the executive should be freed up from running the organization, and certainly from the on-site work, to the degree that he can spend adequate amounts of time outside the organization.

If the supervisor does not spend his time supervising, middle management will have to supervise, because somebody has to see that the work gets accomplished. Now if middle managers are spending their time doing the supervisor's job, then top management is going to have to come down and take care of running the organization, because someone has to make sure that the organization does coordinate and make sure that the work resources are provided. Then top management will be getting involved in budget decisions, getting into standards, getting into work processing. The cost of top management doing these tasks is very heavy—and if top management is doing middle management's work then who is doing top management's work? It is very important that the different levels of management perform those functions that each is designed to perform.

Job description

The Standard Procedures Manual should contain a section devoted to job descriptions. These descriptions should be comprehensive but also as concise as possible. Any job description will require some generalization, but in all cases it must allow the functions of the business to be performed in accordance with the Functional Organization Chart.

In the small business several functional duties must be performed by a limited number of people. This situation does not change the job titles as these are required by the functions being performed. In many cases it is necessary for one person to have many job titles and to perform the various functions necessary to operate the business.

Training employees

The training of employees is one of the most important functions of management at all levels. Proper training requires formal documentation of all matters relating to each job. Without the documentation, the training is apt to become haphazard and to be of little real value.

The Standard Procedures Manual is one basic item of documentation; it is developed and organized to represent the Functional Organization Chart and, therefore, provides an organized means of documenting all the functions of the business.

Some jobs, such as those of the estimator and the job foremen, will require special manuals due to their complex nature. Every employee should be familiar with the personnel sections of the Standard Procedures Manual.

In training a manager it is necessary that the environment of the position being trained for really exists within the company and not some fantasy or theory that does not apply to the position at hand. It is necessary to face reality as to what authority and responsibilities the position actually entails.

The boss is the only one who can give the trainee a real view of the true environment. This is accomplished not by sitting down and telling him about it, but by being an example so that the trainee can see what constitutes success and where the rewards come from. This in-depth training cannot take place if the training is turned over to staff groups or training specialists.

Studies conducted by a large manufacturing company have shown that a person's development is about 90 percent the result of his experience in his daily work; only 10 percent is attributed to special classes, job rotation, educational background, and so forth. Development is the manipulation of the environment and not the person. Development of the person is predicated upon understanding the true environment.

The studies just mentioned also suggested that a large percentage of the development of the managers involved was not due to the fact that they had been to school. The points picked up in school were valuable later but it was the actual performance of those skills which caused them to develop, and the things that really made the difference were (1) that the person was managed in his daily work by his own boss and (2) his relationship with his immediate manager. It was also shown that the thing which helped the person develop more than anything else was the model of management that he worked under.

One of the main responsibilities of managers is to develop those working under them. This cannot be accomplished by sending them to school or special company courses. The prospective manager may very well learn some valuable lessons and discover inadequacies in his own organization, and then come back and try to reform the company; in most cases he will be met with hostility. This could lead to confusion for the manager, and the result may be regression instead of progression. The best development occurs as a result of example.

Managers who are responsible for employee development must realize that it cannot be based on any set of ideal or specified personality traits. Though there are many lists of desirable personality traits for managers, the fact is that there are successful managers who do not fit the picture, and on the other hand there are failures who have all the listed traits.

We should focus on the person's work and its results instead of his personality. Management of the work can then become a separate activity, which we can show the new candidate. Work can be seen, analyzed, and measured, whereas personality is an intangible quality that the manager brings with him. We need to be able to tell the new manager, "This is the type of thing we want you to concentrate on," rather than "This is the sort of person we want you to become."

The effective top manager will insist that opportunity for development be universal. A superficial selection process will result in a very limited number of possibilities or successes, while passing over some of those with apparently lesser abilities who very well might prove to be great managers. The best way to discover talent is to try to give everyone the opportunity to use his talents and skills, because every once in a while we will find a person who is a sleeper. Many workers are in the habit of doing exactly what is expected of them and will do their best to do a good job. But if the stimulus and the opportunity for them to expand their abilities are added you may discover an extremely creative person whose creative ability has been hidden because of his basic attitude.

Most of any employee's development takes place on the job. To develop, a person cannot just be fed the required facts. It is necessary to provide the opportunity and challenge which will be most likely to provide the stimulus to the improvement which we want to take place. Management's responsibility then is to provide the stimulus which will result in the development that they want to take place throughout the organization. Or to rephrase, management development then is primarily a matter of providing opportunity.

Activating the Business

Motivated personnel are necessary to achieve company goals, and the motivation of the personnel is the responsibility of management. Proper supervision and direction must be provided to enable employees to put forth a maximum of productive effort. Internal communications should leave no doubt as to how the desired result is to be obtained.

The preparation and implementation of a documented business plan supported by sufficient employee training is necessary to create a satisfactory business environment and good communications. Regular review and updating of the plan and its documentation is required to assure the progress of the business and high morale among its employees.

So we see that motivated personnel, a documented business plan, and good communications are required to assure the success and stability of the business. And as we noted at the very beginning of this book, it has often been said that management's real job is to build an organization; that then the organization will build the business. *Activation,* consisting of supplying positive controls, imposed by supervision, which prevent departure from the company goals and objectives, is the process necessary for building the business.

Communications

The passing of information from one person to another presents a complicated and often disconcerting problem in the construction industry. It is always difficult for one person to clearly relate detailed information to another person.

In the electrical contracting business, the problem becomes more difficult because many individuals, from the engineering designer to the field installer, must perform functions in a logical sequence and then pass information on to other people. The physical separation of personnel further complicates the process. Due to these factors, communication must be complete, clear, concise, consistent, and simple.

Design standards are the common language used throughout the organization. If correctly used and conscientiously applied, they will allow consistent communications that interrelate all the required functions and operations.

Standards

Since design is fundamental to the entire system and defines the basic methods to be employed in making the detailed installation, it is necessary that the most up-to-date drawings be provided to everyone involved immediately upon execution. Each of the drawings represents a typical construction assembly and provides the means of communication throughout the construction process.

Materials

The primary design of the job determines the pattern to be followed in the field installation. The type of material required for the installation is determined by the type of construction. Research and development in the electrical industry has resulted in many materials which are competitive in price, readily available, and easier to install which can be used to achieve the same result.

Organizational communications

Every electrical contracting job requires the completion of ten basic functions. These are:

1. Design
2. Estimating

3. Sales
4. Processing
5. Purchasing
6. Expediting
7. Warehousing
8. Job management
9. Field installation
10. Accounting

A method of communication must be employed which not only will reach each of these functions but also will provide clear, complete information throughout the progress of the job.

Depending on the preference of management, different companies may have different organizational structures. Regardless of the organizational structure and the size of the company each of the functions illustrated on the Functional Organization Chart must be accomplished. Effective communications depend on the use of a common language which is based on established standards.

Documentation

The best communications are those which are conducted by means of documentation. This eliminates the misunderstandings that often occur in verbal communication and also establishes the what, where, when, and how of a given situation.

Work facilities

The physical environment of the workplace has an important influence on motivation and morale. The poorly designed and furnished office will result in low morale and haphazard work habits of office personnel. The disorderly storage of material and an untidy warehouse operation will result in wasted and lost material along with low morale among the warehouse personnel.

Labor is so costly today that it is necessary to fully utilize each man-hour if a business is to compete successfully. Furnishing the job site with the proper tools, manpower, and material at the proper time are some of the most important functions ensuring the success of the business, and these cannot be accomplished without the proper facilities.

Controlling the Business

It has been said that it is management's responsibility to organize, deputize, and supervise. We have added to these another responsibility: we have said that top management must also fulfill an entrepreneurial function. Now we add one more responsibility to management: that of controlling the business to assure that company policies are being followed and that production in the field is meeting or exceeding budget forecasts.

Before management controls can be established, two conditions must be met:

1. Appropriate budget forecasts must be developed.
2. A simple means must be provided for measuring the results: that is, for measuring how actual performance compares to the forecasts.

The development of these two critical functions is mostly dependent on an efficient accounting department that furnishes the required data on a current basis. The management controls furnished should do the following:

Provide top management with pertinent information on the status of the business at regular periodic intervals.

Provide management with a factual basis for taking prompt action to correct lagging performance.

Provide information which will enable management to fully meet its responsibilities to the business in a manner that will be accepted by all employees while maintaining the highest degree of morale.

Strict controls should be set up for the functions listed below:

1. Accounts receivable aging
2. Billing
   a. Work orders in process
   b. Retainage (amount held until completion of job)
   c. Job suspense account (unbilled work in process)
3. Budgetary controls
   a. Expenditure of funds
   b. Overhead costs
4. Estimating quotas
5. Financial reports and indicators
6. Inventory
7. Job costs
8. Overhead allocation to estimates

9. Overtime worked
10. Profit
11. Purchasing
12. Sales goals and quotas

Methods and reasons for controlling these functions are discussed in the sections that follow.

Accounts receivable aging

Each receivable account must be aged each month according to due date, indicating the amounts due and past due in the following categories:

Current—under 30 days

30 days past due

60 days past due

90 days past due

120 days past due

Over 120 days but within the current year

From last fiscal year

The condition of the current asset known as accounts receivable is determined by the aging record. It is generally agreed that any account that is 90 days or more past due is in perilous condition.

The accounts receivable aging report is readily available each month immediately after the sales journal and cash received journal have been posted to the computer. The aging report will be furnished automatically by the computer without further data being entered. Close control of all accounts receivable amounts enhances available working capital.

Billing

The essential items to control relative to billing are work orders in process, job costs in the job suspense account, and retainage.

Work orders in process. There must be sales entries each month for all items of work done on all jobs during that month. If timely entries of sales are not made, then the expended costs for the same accrue as costs of work done and appear as job suspense (which is unbilled work in process). Thus direct job costs are not billed until there is a corresponding sales entry. To prevent inordinate amounts of working capital being bound up in unbilled suspense accounts it is essential that Work Orders be entered and billed currently, on either a complete or partial payment basis.

Retainage. Retainage often becomes an important factor in financing a business and it is essential to have records which are as accurate as possible. Items necessary to produce these records are (1) accurate labor, material, and job expense costs to date; (2) accurate appraisal of the percentage of completion of the job; (3) accurate prorated approximation of margin (overhead and profit) assignable to the job. An accurate compilation of these items results in a proper appraisal of funds tied up in retainage.

Job suspense account. The account represents unbilled material, labor, and job expense at the end of the accounting period. Usually there are several reasons why these costs have not been billed, such as:

The jobs have just been started; some cost items such as material, bond, or permits have been charged, but no labor has been expended.

The costs are not sufficient to justify billing.

The job is a government job being paid on a percentage of completion basis and costs are being held until the actual sales figure is known to avoid an inaccurate sales and accounts receivable figure.

For good control there should be a definite, acceptable reason why each job having costs against it is not billed each period. Normally, the total of the job suspense account at the end of the accounting period should not be over 4 percent of the net accounts receivable figure.

Budgetary controls

Many successful businesses consider the preparation of a budget in advance of each year necessary for the proper control of income and expense. In some instances, even though the budget is prepared, the business is not successful in controlling the expenditure of funds and overhead costs. To prevent this, it is important that complete control be exercised over each item in the budget. Some budgeted items will have to be adjusted to allow for changes in volume of work. When work in addition to that budgeted and planned for is obtained, provision must be made for additional tools and equipment, and for the increased payroll requirements. Even under these conditions the basic budget should not be changed; that is, the percentage should remain constant.

Estimating quotas

The production of good estimates in sufficient volume to ensure the receipt of enough sales to produce the profit required for the continuing health of the business requires that quotas be established for the estimating department. The electrical contractor who makes certain that he secures a volume of work that generates enough dollars to cover costs and a reasonable profit will be successful. To do this it is necessary to:

Prepare realistic budget projections of the volume of work required for the full year in terms of man-hours.

Allocate the required budgeted man-hours as quotas to the estimator or estimating team.

Maintain a management control register which shows the number of man-hours budgeted versus the man-hours of work awarded. The same register should show the total man-hours estimated and a comparison with a quota of work to be estimated. The types of jobs being bid will affect the comparison, but the register will illustrate to the manager how much work is being processed and by whom.

The suggested register is shown in Fig. 1.3; the use of this form will enable the manager to determine the percent of quota estimated and the percent of quota awarded.

Financial indicators

Please refer to Chap. 3, "Effective Financial Management," where the importance of and the application of the financial reports and ratios is covered completely.

Inventory

The value of controlling inventory lies in the fact that funds tied up in material inventory cannot be applied to essential uses of working capital. With careful planning and the use of computers it is now possible to predict optimum inventory levels and establish procedures and controls to maintain such levels. Excess stock must be returned to vendors for credit.

Overhead allocation to estimates

The control of overhead expense and its allocation as a billable job expense is one of the most important controls in the electrical contract-

Estimating Man-Hour Register

Week no. ____________ Date ____________

Man-hours of work ____________

Department	Name	Man-hours of work estimated						Man-hours of work awarded				
		Quota this week	Estimated this week	Previous estimate	Total to date	Quota to date	% Quota	Quota this week	Quota to date	Awarded this week	Awarded to date	% Quota
Estimating	Peters Jones Avery											
Production	Thomas Mann Rogers											
Sales	Johns Vale											
Line	Thompson											
Service	Tate											
	Totals											

Figure 1.3 Estimating Man-Hour Record.

ing business. For control and allocation of overhead, management needs the following:

- Accurate budget of overhead applicable to each department.
- Accurate forecast of man-hours to be worked by each department.
- Adequate record to indicate the degree of accomplishment of work projected. The Estimating Man-Hour Record may be used as the basis for man-hour information.

Overtime worked

Management may control the working of overtime by:

1. Authorization at the time of working
2. Review of payroll reports
3. Review of time cards

Profit

Profit can only be realized when the number of dollars billed is *more than* enough to offset all overhead costs and direct job costs. Study of the monthly financial statements known as the profit and loss statement (or income and expense statement) and the balance sheet are necessary to control profit. To justify the required amount of billing, it is necessary to reach a maximum level of productivity.

Maximum productivity in the business of electrical contracting can be achieved when the different specialists within the organization pool their collective efforts in order to coordinate estimating, job management, and purchasing with the installation in the field. The result is both that a profit can be made and that the job is completed within the time limits allowed by the owner.

Coordination of the specialists in the business can best be achieved by having a complete business system that permeates the entire structure of the business. This means that all elements of the business must be defined, documented, and transmitted in a form that allows employees to understand and implement the plan of operation.

The efficient installation of an electrical construction job and the achievement of productive man-hours are difficult to accomplish. There are many factors involved, some of which are controllable by the electrical contractor and some of which are not. The planning, scheduling, and mastery of the details of installation by everyone concerned are mandatory if satisfactory results are to be achieved, and this demands close coordination between those developing the estimating data, the job management, and the field personnel who make the installation.

Purchasing

Using the timely reports now available as generated by modern day computers it is now possible to compare estimated prices with material prices paid for every item of material on the job.

Sales quotas

Salesmen must have incentives to function properly. Setting a required volume of sales provides one incentive. Properly compensating the person who achieves it provides another. Both are necessary.

Sales quotas are also fundamental to budgeting; after they have been established, it is a clerical operation to compare monthly the quota with the actual accomplishment.

Personnel Management

The manager must realize that people are the company's greatest asset and help the people who work in the organization to pursue their own interests, and to fulfill their needs and aspirations. This interest in their personal well-being must be expressed not only through tangible means like wages, benefit plans, and the availability of programs for self-development and training, but also by instigating the cordial and constructive working and leisure relationships that will encourage people to work as a team.

Wage evaluation

To many employees the most important feature of their job is the compensation. The actual wage may be only part of the compensation package, but it is the factor that determines the individual's standard of living. Therefore, it is important that management protect the employees' interests by establishing a wage policy that is fair to all concerned, in balance with the economy, and not detrimental to employee morale. To do this it is necessary to establish a base wage for each position and set standards to be used in evaluating employee performance.

The base wage for each position in the company should be determined by a study of the community standard wage for comparable positions and, if possible, the base wage your competitors are paying for this position. Most of the wage studies conducted by the states or local communities will indicate a low, median, and high wage for each job category. It must be remembered that these studies include input from a broad spectrum of businesses and, therefore, probably do not provide an accurate standard for the electrical contracting business. Because of this, the wages noted in the various studies should be used only as a guideline in establishing your base wage rates.

There are five basic elements to be considered when evaluating an employee's job performance; these are:

1. *Quality:* Ability to work to plans and/or specifications
2. *Reliability:* Compliance with company policies, procedures, attendance, promptness, conventional appearance, personal hygiene, safe work habits
3. *Quantity:* Productivity level of acceptable quality
4. *Attitude:* Willingness to comply, cooperate, get along with supervision and fellow workers
5. *Job knowledge:* Knowledge, experience, continuing education, skills levels

The five elements may be used to develop an evaluation form such as that shown in Fig. 1.4. Such a form greatly simplifies the evaluation process for the manager or supervisor. When the form is completed it should become a permanent part of the employee's personnel file and be referred to each time an evaluation is performed.

The evaluations should be undertaken for each employee at regularly scheduled times at least once a year. It is extremely important that the results of the evaluation be discussed with the employee and that the employee be advised of the reasons for the points assigned to each item and also what the employee can do to improve the score in future evaluations.

Company policies for personnel conduct

All company rules and policies regarding personnel should be in writing, and every employee should have a copy readily available. The policy of full distribution of personnel policy avoids many questions and possible disputes and simply results in better management. A sample policy statement follows.

Employee name ______________ Job title ______________ Date ___/___/___

Points	20	15	10	5	1	Reasons	Points
	Descriptive point Levels						
Factor	Ability to work to plans and/or specifications						
Quality	Excellent: Always works to spec. & under difficult conditions	Good: Rarely does poor work	Fair: Occasionally does poor work	Poor: Frequently does poor work	Very Poor: Very often does poor work		
	Compliance with company policies, procedures, attendance, promptness, conventional appearance, personal hygiene, safe work habits						
Reliability	Excellent: Always in compliance with all requirements	Good: Rarely fails to comply	Fair: Normal compliance	Unreliable: Often undependable	Very Unreliable: Completely undependable		
	Produces to standard labor unit requirement						
Quantity	Excellent: Consistently exceeds standard	Good: Frequently exceeds standard	Fair: Occasionally meets standard	Poor: Seldom produces at standard rate	Very Poor: Consistently turns out very little work		
	Willingness to comply, cooperate, get along with supervisers and fellow workers						
Attitude	Excellent: Always complies with objectives gets along with other workers	Good: Usually complies, cooperates, gets along with others	Fair: Occasionally complies and has trouble with others	Poor: Frequently reacts begrudgingly does not get along well	Very Poor: Generally un-cooperative and difficult to work with		
	Knowledge, experience, continuing education, skills level						
Job knowledge	Excellent: Performs perfectly with little assistance	Good: Performs most aspects of job and rarely requires assistance	Fair: Quite often requires assistance on complex operations	Low: Performs limited part of job or requires close supervision	Very low: Fairly unskilled and requires close supervision		

Evaluated by ______________ Job title ________ Department __________ TOTAL POINTS __________

Date for next review ___/___/___

Figure 1.4 Employee Wage Evaluation.

PERSONNEL POLICY OF ABC ELECTRIC

WORK HOURS

Office The office will be open from 8 A.M. to 5 P.M. daily Monday through Friday. The Lunch period is 12 noon to 1 P.M. with the exception of the receptionist who will take Lunch from 1 P.M. to 2 P.M. This staggering of office lunch hours is to allow a regular office employee to act as receptionist during the 1 P.M. to 2 P.M. period.

Service Department The service dispatcher works from 7 A.M. to 3:30 P.M. Monday through Friday. The lunch period is one-half hour from 12 noon to 12:30 P.M.

Shop Mechanic The shop mechanic works from 7 A.M. to 3:30 P.M. Monday through Friday. The lunch period is one-half hour from 11:30 A.M. to 12 noon.

Warehouse Personnel The warehouse personnel work from 7 A.M. to 3:30 P.M. Monday through Friday. Lunch period is one-half hour from 12:30 P.M. to 1 P.M.

Field Personnel All field personnel work according to the requirements of their jobs. Their work week ends on Wednesday night. The paychecks will be distributed on Friday at the job site by the general foreman. All time cards are to be filled out completely, showing job number, system number, and if applicable the job division number and the hours worked each day on each number.

TIME REPORTS

All hourly employees are required to fill out time cards. Each day's time should be filled in at the end of the day. The report shall be approved by supervision who will then turn it in to the accounting department.

OVERTIME

All time over 40 hours per week worked by nonexempt employees will be paid at time and one-half. Sunday and holiday time will be paid at double time.

VACATIONS

Permanent full-time employees will earn one week paid vacation after the first six months of service. Service time is computed from employee's starting date. After the first six months vacation will accrue at the rate of one day for each six months of service in each year or two weeks per year, for the first five years of service. After five years of service all employees will be entitled to three weeks vacation per year.

All employees are entitled to vacation time and are expected to take earned vacation within the current year. No salary will be paid in lieu of vacation time.

SICKNESS AND DISABILITY BENEFITS

Sickness and/or disability benefits will accrue from the first day of service at the rate of one day per month or twelve days per year of service not to exceed thirty days. Sickness and/or disability benefits are not payable in duplicate with compensation from other sources of insurance or workers' compensation.

Upon termination of employment unused sickness and disability benefits are not reimbursable.

(Continued)

HEALTH INSURANCE

The company provides health insurance with a national carrier as an added benefit for employees. At the end of the first month of employment, the employee is eligible for coverage. Employees may add their dependents to this coverage and pay for their share through payroll deductions if they choose to do so.

LIFE INSURANCE

A $2,000.00 life insurance policy is also provided for employees by the company as a part of the health insurance plan.

HOLIDAYS

Nine legal holidays are observed by the company: New Year's Day, Memorial Day, Veterans Day, the Fourth of July, Labor Day, Thanksgiving and the day following, Christmas eve, and Christmas day. When these holidays fall on a regular work day, straight-time pay will be paid.

DEDUCTIONS FROM PAY

Required legal deductions such as FICA and Withholding Tax are made from each payroll check. Other deductions such as savings and charitable deductions will be made at the request of employees.

JURY DUTY

Employees will be paid up to two weeks salary while serving jury duty. It is required that jury fees received during this period be reimbursed to the company. The employee is expected to spend free time during the period of jury duty on the job.

BONDING OF EMPLOYEES

All employees are automatically placed under a $10,000 blanket position bond upon being employed with the company. At the time of employment, each employee is asked to fill out an application blank for the bonding company's information, investigation, and records.

EMPLOYEE SUGGESTIONS

Employees are encouraged to offer suggestions on ways and means of improving methods and procedures, and the company requests that employees be alert to possible improvements. A suggestion box is fastened on the wall next to the bulletin board.

LUNCH ROOM

The kitchen and training room are available for use by any employee who wishes to bring their food and drink for lunch. Employees are responsible for keeping both areas clean.

COMPANY INFORMATION AND DATA

All company forms, data, catalogs, and business procedures are confidential and must not be revealed to persons outside the company for any reason.

(Continued)

COMPANY INFORMATION AND DATA (Cont.)

No information relating to a bid or quotation in any form should be transmitted to outsiders.

Patents, copyrights, etc. developed while the employee is in the employment of the company are considered company property.

SALARIES

The management of the company will review all salaries at least once each year and evaluate each salary according to a point system. Each employee will be advised of the rating assigned to them and the reasons therefore. Suggestions will be made by the reviewing manager as to how the employee might improve their rating for the next evaluation. It is the intention of the company to maintain an above-average salary and wage scale commensurate with the particular job.

OUTSIDE WORK

Employees are discouraged from performing outside work which is in any way similar to that performed by the company.

PROFIT-SHARING PLAN

The company has a profit-sharing plan which was established to provide financial security for the employees' retirement.

Every full-time employee who is not subject to mandatory bargaining and has completed one year of service with the company is eligible to be a participant in the plan. Participation in the plan will begin on the first day of the month following the date of completion of one year of service.

The company will contribute the maximum allowed by law to the plan limited only by the amount of profits accrued by the company during the applicable accounting period. At the present time the legal contribution limit is 15 percent of salaries paid during the current year.

Each employee's share will become vested in him at the rate of 10 percent per year of service. At the end of ten years of service all contributions made to the plan in the employee's name before that time, and thereafter, will belong to him subject to the legal provisions of the plan.

Additional details about the profit-sharing plan are available from management and the Profit-Sharing Plan Booklet.

PERSONAL TELEPHONE CALLS

Personal telephone calls are discouraged but will be allowed on a limited basis if the call is absolutely necessary. All personal calls should be held to the minimum. Employees should advise callers of company rules so that they may be guided accordingly. Personal long distance calls are not permitted on company telephones.

LATENESS

Customers expect our company to be open for business and available for services during our regular hours. Employees who are repeatedly late show a disregard for company rules and for others who must carry on and therefore this practice will not be tolerated.

(*Continued*)

ABSENCE

An employee who is unable to report for work because of illness, or for other good reasons, should call his department head as soon as possible after 8 A.M.

POLITENESS AND COURTESY

It is important that employees be polite and courteous at all times, not only to customers but also to other employees. Full cooperation between departments is essential.

PERSONAL APPEARANCE

Customers form an opinion of a business based on the appearance of its employees; therefore, it is important that all employees be as neat and well-groomed as possible at all times.

OFFICE HOUSEKEEPING

Offices and desks should be kept as neat as possible. Desks must be cleared off at night so that the janitor can dust them.

Surplus paper and pencils should be returned to the accounting department for storage.

Company forms should not be used for scratch paper.

All office machines should be covered after office hours.

CORRESPONDENCE

All letters should be produced on the word-processing system. Company letterheads should not be used for personal correspondence.

INTEROFFICE COMMUNICATIONS

Interoffice visits should be made only when necessary and the telephone cannot accomplish the purpose of the visit.

MEMOS

It is important that messages concerning requests, information, or instructions to another person or department be in written memo form. This is to ensure that the message is understood and not forgotten.

SAFETY

All accidents must be reported to the safety director immediately so that proper action may be taken to obtain treatment of any injury and to report the accident to the company insurance carrier.

The Manager of the Future

Throughout this book, and particularly in this chapter on effective business management, we stress the importance of efficiency and of the effective control of a business as being necessary to produce a profit and thus enable a business to survive and prosper. One of the authors, in a presentation to the NECA Academy of Electrical Contractors, defined *efficiency* as "effective operation as measured by a comparison of production with cost. It appears that this is the crux of the whole matter of productivity—to be able to bring about effective production within the realm of competitive cost limitations. Performance such as this implies 'excellence in operation.'"

The tremendous advances in technology over the past decade and the complexity of modern society have created an often baffling environment, in which leaders and management are compelled to face forces of change which have never existed before. Competent management has always had to look at the future and determine which situations would or could have an effect on their business, but the current situation demands an even more astute look than was required a few years ago.

For example, the U.S. Department of Labor has predicted that there will be a severe shortage of skilled labor during the next two decades and that businesses are going to be faced with the problem of training workers themselves because existing training facilities will be unable to provide skilled workers in anywhere near the numbers required. The problem is that many of the available workers will be not only totally incompetent but illiterate, and, therefore, extremely difficult to train.

The successful manager of the future must prepare now to have a competent labor force available to the company to be able to compete in the marketplace as it will exist in the near future.

Management Style

The effective manager must develop a style or theory which works for him; this theory will be most effective when it is a simple, sensible one, based on the concept that people and the quality thereof are more important than the type of organizational structure. Some suggestions and guidelines for this type of management style are:

Simplicity: Avoid too complex an organizational structure; such a structure only makes it difficult to achieve simple, direct communication. Avoid complicated compensation schemes and keep reward and incentive devices simple and easy to understand. Make one-page memos and reports do the job.

Integrity: Maintain the highest ethical standards. Produce only to the highest standards of performance and reliability.

Quality: Lead the industry in providing the highest-quality and most efficient, competitive, service possible.

Entrepreneurship: Take calculated risks and encourage creative thought in employees. Reward employees for good ideas.

Adaptability: Be alert to changing conditions in the marketplace which affect your business or which may affect it in the future.

Respect and openness: Encourage employees to be open, honest, and fair with every other employee. Communication must be allowed to flow freely throughout the organization.

Concern: People are management's greatest asset. Help your employees to reach their greatest potential and in addition to contributing to their personal well-being through tangible means, encourage constructive and cordial working relationships.

Fostering of company community: The involvement on the job is a major portion of the employee's life. Make it clear to all employees that the work of each affects not only his or her own well-being but also the well-being of every other member of the organization.

Encouragement of teamwork: Encourage a team or group approach to problem solving. Most production is accomplished through teamwork. The sharing of ideas and responsibility often leads to higher productivity and greater personal fulfillment.

Striving for excellence: Effective management must have an ongoing commitment to the attainment of excellence. It must lead, not follow. *Excellence* means the outstanding good of its kind and the best that can be produced by human inventiveness working under the most favorable conditions that can be arranged.

Planning for growth: Effective management must always plan for growth and never accept any philosophy of limits to growth. It must plan with foresight and act with ordered determination, and growth will be within reach.

Chapter 2

Effective Use of the Computer

Mechanical computers date back more than 5000 years to the invention of the abacus, which probably occurred in China. The abacus consists of a wooden frame with wires that are strung with beads. The beads represent units: tens, hundreds, thousands, and so on. Calculations are accomplished by moving the beads up and down. More than 2 million abacuses were being manufactured each year in Japan as recently as 1983. The abacus is still used in the Far East and the former Soviet Union.

Adding machines and early computers remained strictly mechanical devices until 1930 when the first electromechanical analog computer was developed. In 1942 an electronic digital calculator was constructed at Iowa State University. The first fully electronic computer, built at the University of Pennsylvania, followed in 1946.

Since 1946 the development of the computer has advanced at a rapid pace and during 1975 the first personal and affordable computers began to appear. The introduction of these machines, commonly referred to as "personal computers" (PCs), launched a revolution that brought the computer into homes and schools, as well as into every aspect of business.

The introduction of the PC solved the problem of large amounts of space required for the mainframe and mini computers manufactured prior to 1975; the PC is a compact, desktop unit. But while the physical size has been reduced the capacity to perform has been steadily increased, until today most electrical contractors will find the PC more than adequate to handle all their requirements.

The PC is particularly adapted to the electrical contracting industry and is being used in general accounting; it can be used to maintain all the accounting journals, such as those for sales, accounts receivable, accounts payable, and payroll. Computers are also serving a valuable function in purchasing, inventory control, budget development and control, cost accounting, tool control, and estimating. Many contractors are placing PCs on the job site to aid in producing Change Orders, in labor control, and in material control.

One of the most valuable contributions made to the industry by the PC is in the important field of estimating. Timely reports to management are another valuable function of the computer.

Contractors today need not fear encountering the difficulties experienced with computing machines before the development of the PC, either in developing a new system or advancing to computers from an obsolete system. The computer industry has solved the problems encountered with the computing machines of a few years ago.

Installation of a modern computing system does require astute management, proper planning, purchase of a viable system, and good programming. The advantages obtainable through the use of computers as discussed in this chapter are certainly available to all, but require the active involvement of management in the acquisition and application of the system. It is vital to the success of the computer operation that management know what information it requires for the management and successful operation of the business. Then it is necessary for management to ascertain that these requirements are satisfied, not only in the acquisition of the hardware, but also in the development of the programming. Too often in the recent past management has assigned the responsibility for these requirements to a computer salesperson, a programmer, or a company employee who was not really aware of the total picture.

It is also imperative that management ascertain that steps to implement the system are taken in the proper order. Planning should determine which operations of the business will utilize the computer and the timing and order of installation. For example, the estimating program should be up and running before attempting to install a purchasing and budget program because the purchasing and budget programs will utilize information available from the estimating operation.

The employee "attitude problem" regarding computers in the past, and to some extent at the present time, is simply a result of resistance to change. Generally, this problem is encountered among older employees who have never been exposed to computers and in many instances are not system oriented, and cannot adjust to the strict disciplines required in using a computer. Most people entering the work force in the past 10 or so years have been somewhat exposed to, and are anxious to use, the computer. Some of the older members of the work force will continue to resist change and in some cases will never accept the computer. These people will always find some reason why the computer will not work, and they may have to be replaced or transferred to another position where use of the computer is not required.

To compete in the industry today, a contractor will find that the computer has become a necessity. Economics alone require the use of a computer in estimating, if for no other reason than to control the cost of preparing the estimate. One small computer, by eliminating much of the paperwork, utilizing manual unit pricing, manual application of labor units, and then computerized extensions to determine the estimated cost, can increase the number of accurate estimates produced by the same number of estimators by as much as 30 percent.

In addition to the economics, the computer reduces the margin of error in material pricing and application of labor units and costing. All the industry material pricing catalogs are now available in computer format for installation on the contractor's computer.

The computer has no equal in all phases of the accounting function. Payroll and the required union reports, tax returns, application to job cost, audit trail, and management reports are handled with one entry on the computer.

All phases of and the various types of billing, such as time and material billing, the various forms of cost plus, or regular percent-of-completion progress billing have become simple operations on the computer. Labor costs accrue directly to each job as a direct distribution from the payroll. Material costs can be directly costed to billing from Accounts Payable in the same manner. Overhead and profit percentages are automatically applied.

Purchasing budgets and controls are developed from information contained in the estimate. Expediting and inventory can be controlled on a daily basis, thereby eliminating overstocking. Accounts payable and cash flow projections are a side benefit of the reports.

Timely financial statements are virtually an automatic function of other programs, and are available whenever needed. Comparisons of vital information from previous periods, such as the statement of profit and loss or the balance sheet from the previous month or the previous year, are also available.

Estimating

The fact that estimating is the first step toward making a sale establishes it as one of the most important functions in the electrical contracting business. The computer establishes a standardized, orderly, efficient, and accurate means of estimating.

In addition to improving the efficiency and accuracy of the estimate, the computer generates several valuable reports. These include a detailed Bill of Material by divisions of the estimate, and a consolidated Bill of Material for the entire estimate, listing all the material by type, which can be used in obtaining quotes from vendors and checking available on-hand inventory to determine what material does not have to be purchased. Total labor requirements, as well as purchasing budget guidelines, become available from the information generated in the basic estimate.

At the same time, the computer is flexible and allows the estimator to apply personal knowledge and required variables such as job factors, problems in installation or scheduling, or any other items which would affect the cost of performing the job.

To utilize a PC-type computer for estimating, one of the database programs or one of the many estimating systems on the market is necessary. The data entered into the program would be the standard labor units established for your company with appropriate code numbers assigned, the catalog material prices, and a range of discounts available to your company from your vendors. The database program has the capability to apply the labor unit and material prices to your code numbers. Refer to Fig. 2.1 and Fig. 2.2 for an example of the computer entry screen and computer-based database to accomplish pricing and extensions.

The program will then price and extend the individual line items, which may then be printed and used for checking the input. Based on the hypothetical estimate, material prices, and labor units in Figs. 2.1 and 2.2, the computer would print out an estimate for material and labor as:

Code no.	Quantity	Description	Cost	Labor hours
10001	200′	1/2 EMT	$100.00	8.8
10002	100′	3/4 EMT	60.00	4.5
40015	25	4/s Box	18.75	2.5
40025	50	4/O Box	37.50	5.0
50030	440′	14 THHN	88.00	12.3
50035	210′	12 THHN	52.50	6.1
Total material cost and labor hours			$356.75	39.2

The database program also performs the sorting of the information entered, necessary to assemble the information in the desired order for the required reports. Several of these programs, such as R-Base or D-Base, are available at reasonable cost.

There are also canned or prewritten estimating programs on the market. If you have an established estimating system, it is probably best to use one of the database programs and have your program cus-

Estimate no. 102 Job no. P31492 System 3 System name ____

Branch ____

Section no. ____ Estimated by RJJ Date 01/03/92

Job name J. Warehouse Column A

Code no.	Quantity	Unit	Factor, %
10001	200	100/ft	+10
10002	100	100/ft	+10
40015	25	ea	
40025	50	ea	
50030	440	100/ft	
50035	210	100/ft	

And so forth.

Figure 2.1 Entry screen to access database for pricing and extensions of the estimate.

Code no.	Description	Unit	Material price A	B	C	Labor units A	B	C
10001	1/2" EMT	100/ft	$50.00	$55.00	$60.00	4.0	4.4	5.0
10002	3/4" EMT	100/ft	60.00	66.00	72.00	4.1	4.5	5.4
40015	4/s Box	ea	.75	.90	1.00	.1	.2	.3
40025	4/0 Box	ea	.75	.90	1.00	.1	.2	.3
50030	14 THHN	100/ft	20.00	22.00	30.00	2.8	3.2	4.0
50035	12 THHN	100/ft	25.00	28.00	35.00	2.9	3.3	4.1

And so forth.

Figure 2.2 Sample database for pricing and extension of estimates.

tomized to your specifications. This basic program can also be utilized in your operation for other computer functions, such as all the various accounting functions.

The five primary functions of estimating are:

1. Registration
2. Takeoff
3. Tabulating
4. Pricing
5. Summarizing

Each of these functions is readily handled by a computer.

The information required for registering the job and estimate is quite simple to enter on the computer using any of the standard database programs. The form which may be used is shown in Fig. 2.3. The form and the information may be changed to adapt to your operation, but it is suggested that most of the information be retained as it has proved to be important for use of management in future bidding and possible perusal of claims.

In addition to the speed advantage the computer has over the handwritten form, the computer form can be "flagged" to require some of the information to be entered before the form will leave the screen. Also, the form can remain available in the computer for instant use until the time when the job is awarded. Once the necessary copies of the form are printed and filed, the information on the form is cleared to allow entry of the information on the next estimate. It is important to note that the job name and number assigned to the estimate here will be used throughout the life of the job.

The computer estimating system does not eliminate the requirement of making subdivisions by system or division of the estimate, designing the installation, or selecting material.

Before beginning the takeoff, the plans should be categorized by job system. This will allow several estimators to work on the takeoff at the same time, accelerating its completion. A 10-item list of job systems could be set up as follows:

1. Service entrance, metering, and grounding
2. Lighting fixtures
3. Distribution equipment
4. Raceway equipment
5. Feeder and busway
6. Branch circuit
7. Outlet detail
8. Power systems
9. Telephone and communications systems
10. Special systems

The computer can be programmed to accept the takeoff information either from a manual takeoff or through use of the electronic probe counter and measuring device. Whichever takeoff method is used, the computer will require a code number for each item of material required. In most cases this code number will be 5 or 6 digits and will identify either one item of material or an assembly requiring several

Job Registration

Estimate no. ____________________

Estimator ____________________

Job name ________________________________ Date ____________________

Job address ________________________________ City ____________________

Architect ________________________________ Contact ____________________

Engineering firm ________________________________ Contact ____________________

Electrical engineer ________________________________ Contact ____________________

Bid closing date ______ Time ______ Depository closing date ______ Time ______

Duration of job ______________ Estimated man-hours ____________

Plans obtained from ________________ Return date ____________________

Plans picked up by ________________ Plans returned by ____________________

Deposit required ______ Deposit returned date ______ Returned to ____________

Special instructions to estimator ________________________________

____

Information on results of bidding:

Electrical contractor	Base price	Alt 1	Alt 2	Alt 3	Alt 4

And so forth.

Electrical contract awarded to: ________________________________

Information on results of general contractor bidding:

General contractor	Base price	Alt 1	Alt 2	Alt 3	Alt 4

And so forth.

General contract awarded to: ________________________________

Notes on bidding ________________________________

Total man-hours electrical ____________ Maximum no. electricians ____________

Figure 2.3 Job registration.

items of material such as would be required for an outlet. The ability to enter a code for an assembly reduces the number of keystrokes necessary for entry of the individual items of material.

The code numbers used in your database to identify the various materials should be set up in a logical sequence. For example:

Code no.	Unit	Material	Price	Labor
10001	100′	1/2-in EMT	$65.00/C	3.0
10002	100′	3/4-in EMT	75.00/C	4.0
10003	100′	1-in EMT	82.00/C	4.6

When the manual method is used, it is necessary to apply the code numbers to each item before the computer entry can be made. When using the computerized method, the code number of the item being counted must be entered on a keyboard before the count begins.

Regardless of which system is used, manual or computerized, it is mandatory that the computer entry be checked against the original takeoff quantities and the code numbers. The computerized method requires proofreading, while the manual verification can be accomplished by running an adding-machine tape of quantities and another tape of code numbers. The computer will automatically give the correct total of these two items, which can then be compared to the totals on the adder tape. If the totals agree, there is very little chance of error in the entry.

The coding of the estimate should be done by the estimator, but the computer entry should ideally be done by a computer entry person other than the estimator. This supports the estimator and allows him to better utilize his time.

The job name, job system name or job section, date of the estimate, and date of entry are entered in the computer by the entry person before entering the coded takeoff information. (Refer to the entry screen in Fig. 2.1.) Once the code numbers and quantities have been entered, the computer will do all the necessary calculations for material quantities, extended prices, and application of labor units and labor dollars. A detailed priced and extended printout of each system or division of the estimate is available as it is entered. The printout will include all the identification information entered on the entry screen and will also show the estimate figures, as illustrated on the hypothetical estimate at the bottom of Fig. 2.2.

Upon receiving the printout for each system or section of the job the estimator should examine the plans and determine if all possible deviations from normal installation practices required on this particular section of the job have been taken into account during the takeoff. He should also check to make sure that all factoring he has determined to be necessary has been properly accounted for during the computer entry process. If not, he should indicate the percentage either plus or minus applicable to any of the items and return the printout to the input person, who will enter the factor indicated in the proper column and then recalculate and supply the estimator with a corrected printout of that system or section of the estimate.

When all adjustments to the estimate have been made, the computer can tabulate all the entries and furnish a consolidated Bill of Material printout, priced and extended with the standard and factored labor and material prices for the use of the estimator. Another consolidated listing showing only quantities and material description can then be printed out for use of the vendor in quoting material prices.

When all quoted prices and variables have been entered, the next step to be performed by the computer is to assemble and make a summary recapitulation sheet of the estimate. The computer will apply the cost of overhead, tools, supervision, rental equipment, any other miscellaneous items, and profit to the estimated costs and arrive at a selling price. Refer to Fig. 2.4 for an illustration of the computerized estimate summary.

All the above information will remain in the computer until the contractor is notified of the bid results. If he is awarded the job, the computer will then use estimate information to prepare a Bill of Material and to set up purchasing schedules conforming to the job schedule and the lead time required for ordering the material. The purchasing schedules are then used to set up budget and cash flow requirements for the purchase of material.

In addition, the computerized Bill of Material becomes a checklist, used to ascertain that all material has been ordered and when to expedite delivery of ordered material.

The estimate information is then used by the computer to total the labor requirements by labor code for the individual systems and/or sections of the job, and also the totals for the entire job. These figures allow the contractor to calculate his labor requirements according to the critical path furnished by the general contractor to establish job progress.

The functions noted above can be programmed into the estimating system. However, if the contractor has been operating on a manual system, it may be advantageous to have the program follow somewhat the same task order as the manual system. This becomes a management prerogative and a factor in the acceptance of the computer by rank and file employees.

The computer offers the ultimate in flexibility in pricing the estimate. The material price column and the labor unit column in the da-

Estimate Summary

Estimate no. ________

Job name ______________________________ Date __________

Job location ____________________________

Bid depository no. __________ Closing time ______ Filing time ________

Estimated by ____________ Checked by ______ Approved by ________

Sheet no.	Section of job	Std. materials	Spec. materials	Labor (man-hrs)
Subtotals				

Direct job exp.	Job factor, %	Subtotal std. mat.
Bid bond ________	Type building ________	Misc. mat. & labor ______
Insurance ________	Work cond. ________	Labor job factor, b ______
Insp. fee & lic. ________	Gen. contr. ________	NPL., c ________
Gen. super. ________	Elec. contr. ________	Labor prod. factor ______
Eng. & Dwgs. ________	TTL JOB FACT, % ______	Total mat. & lab. ______
Storage & job ofc. ______	% NPL, man-hrs ______	Ave. labor rate per hour $ ________
Timekeeper ________	Foreman super. ______	Total $ value labor ______
Freight & exp. ________	Study time ________	Total value material ____
Truck exp. ________	Mat. handle ________	Total value direct job exp., d ________
Telephone ________	Order mat. ________	TOTAL PRIME COST ________

Figure 2.4 Computerized estimate summary.

Estimate Summary

Temporary power ______	Testing ______	OVERHEAD (based on cost), % ______
Tools & equip. ______	Lost time ______	OH-Std. Mat.-Lab.-Job. Exp., % ______
Travel exp., days @ ______	TOTAL NPL, c ______	OH-Spec. Mat. & Equip., % ______
Travel exp., mi @ ______	LABOR, % ADDER ______	TOTAL GROSS COST ______
Rm. & bd., days @ ______	FICA ______	PROFIT, % ______
Labor adder, % ______	SUTA & FUTA ______	SUBTOTAL ______
______	Workmens comp. ______	Sales tax ______
______	PL & PD ______	Excise tax ______
______	Assoc. dues ______	Payment & performance bond ______
______	NEBF ______	Interest on financing ______
______	Local health ______	TOTAL ESTIMATED PRICE ______
TOTAL JOB EXP., d ______	TOTAL ADDER, % ______	AMOUNT OF BID ______

NPL = nonproductive labor.
TTL = total.

Figure 2.4 *(Continued)*

tabase should actually consist of three or more columns each, labeled A, B, C, etc.

Column A under Material would contain the standard pricing for bulk material purchases and column A under Labor would contain the standard new construction labor units. The "B" columns would show pricing and labor units applicable to Change-Order work, which on most jobs are considerably higher than the standard prices and labor units, because most Change-Order work involves remodel-type working conditions. The "C" columns would contain prices and labor units reflecting plus or minus percentages for add-on or deducts to allow for special discounts available, various job factors developed by the estimator during the takeoff, or other special conditions which apply to the particular job or time frame.

Other columns may be added to fulfill the individual company's pricing and labor requirements. For example, the contractor may need to allow for the unavailability of skilled labor and the necessity of using more semiskilled labor. This situation would increase the required labor units, since the use of semiskilled or partially trained labor in many cases will effectively double the installation time. Special pricing of some items of material by the supplier could result in either higher or lower material costs.

Accounting

The computer has become an absolute necessity in the accounting function of most businesses. All four of the standard accounting journals—accounts payable, sales, accounts receivable, and payroll—are handled beautifully by today's computer with an extremely high degree of accuracy and efficiency. All the information necessary for the proper operation of the business that is obtainable from the journals is readily accessible daily, weekly, monthly, and annually.

The quality, quantity, and usability of the information obtained from the computerized system is hardly even comparable to the old manual bookkeeping systems. With labor usage reports available weekly on the job site the company management and job management is able to tightly control job labor and identify problem areas almost immediately. The payroll information may be furnished, for each job, on a form entitled Weekly Payroll Report as shown in Fig. 2.5, and all the required current information is obtained automatically by the computer from the current

Weekly Payroll Report

Week Ending _ _ _

Job no. P103091 Job system no. ________

Labor code no.	Total labor hours to date	Total labor dollars to date	Total estimated labor hours	Total estimated labor dollars	To date over/under estimated	
					Hours	Dollars
2105	40.5	$ 648.00	210.0	$3,360.00	-169.5	-2712.00
2110	200.0	3,200.00	150.0	2,400.00	50.0	800.00
2530	150.0	2,400.00	150.0	2,400.00	0	0
3410	350.0	5,600.00	400.0	6,400.00	-50.0	-800.00

Figure 2.5 Weekly Payroll Report.

payroll and by accessing job cost records for previous payroll periods by job number to obtain the accumulated totals.

Accounts payable

An Accounts Payable Vendor Master containing the information shown in Fig. 2.6 should be set up in the computer. Entry of the vendor ID number when posting to the Accounts Payable input screen will automatically create a vendor's name entry on the Accounts Payable Invoice Register. Figure 2.7 illustrates the input screen used to input vendor invoice information to the computer. All the invoice information necessary to complete the Accounts Payable Invoice Register will be transferred from the input screen automatically by the computer. The invoice amounts accumulate as the year-to-date (YTD) purchase amount on the Accounts Payable Vendor Master.

Accounts Payable
Vendor Master

1. Vendor ID no. ____________ 2. Vendor name ________________________

3. Remit address ________________________

____

____

6. Contact name

7. Telephone (____) ____ -________ 8. Date estab. ___ ___ ___

9. Last active ___ ___ ___ 10. Our acct. no. ________

11. YTD purchase ____________

Display Next Previous Enter Change Remove Help Quit

Figure 2.6 Computer setup for Accounts Payable Vendor Master.

Vendor no. ________ Vendor invoice no. ____________ Invoice date __/__/__

Invoice due date __/__/__ Invoice terms ______ Discount, %____

Job no. ________ Job system no.________ Invoice amount $ ______ . ____

General ledger account no. _____ . __

Figure 2.7 Computer entry screen for Accounts Payable Invoice Register.

The Accounts Payable Invoice Register (see Fig. 2.8) should contain the following information:

1. Vendor or supplier name and ID or code number
2. Vendor or supplier invoice number
3. Invoice date
4. Invoice due date
5. Job number
6. Job system number
7. General ledger account number
8. Invoice amount
9. Discount amounts available

All this information will be used for valuable reports during and after the current accounting period.

It is advantageous to list the vendor by a code number rather than the full name because of the keystrokes saved in inputting the information. The vendor's code number inputted on the input screen will enable the computer to write the check to the proper vendor when the invoice is processed for payment, apply the invoice amount to job cost by job number, and post the invoice amount to the correct general ledger account.

At the end of any given period, the computer, by sorting all payments made to this vendor, and then adding the current total to the previous period totals, will determine the total amount of business the particular vendor has received from the company for the year-to-date period. This information will be posted to the vendor master automatically each time invoices are entered. In some instances this information may be used in negotiations for better service or prices. At other times it can be used to determine a more even distribution of purchase orders to a group of suppliers.

The vendor or supplier's invoice number is required to establish an audit trail of receipts of material or services and payments made on all invoices entered into the system. This number will be entered by the computer on the voucher of each check written to the supplier and used to verify payment.

The invoice date and the invoice due date will be used to further identify the invoice and establish the company's payment date. The due date of the invoice will be used to determine the company's required cash flow for any given period.

The "job number" entry allows the computer to distribute the cost

Vendor no.	Vendor name	Invoice no.	Inv. date	Due date	Inv. amt.	Disc. amt.	Job no.	Job sys. no.	GL acct. no.
1001	ABC SUPPLY	2465	03/10/91	04/10/91	$45,623.00	$456.23	P31090	P31090-3	120.1
1125	B&J SUPPLY	69801	03/07/91	04/10/91	545.30	10.90	P20791	P20791-1	120.1
1746	GREG SUPPLY	14207	03/26/91	04/25/91	1,234.54	24.69	P21091		120.1
1803	HAPS MECHANICAL	1004	03/03/91	04/10/91	225.46	2.25	STOCK		121.0
2130	J&J ELECTRIC SUPP	2968	03/17/91	04/10/91	2,540.91	50.81	P30191	P30191-6	120.1
2645	NOW SUPPLY CO.	24356	03/04/91	04/10/91	1,567.32	15.67	P21091		120.1
2780	OWL OFFICE EQUIP	18902	03/08/91	04/10/91	104.98	1.05			525.0
2780	OWL OFFICE EQUIP	18945	03/12/91	04/10/91	35.00	0.35			525.0
2845	PERFECT ELECTRIC	32654	03/02/91	04/10/91	5,678.00	$113.56	P20791	P20791-1	120.1
4545	WALTS ELECTRIC	1587	03/12/91	04/10/91	254.95		STOCK		121.0

And so forth.

Figure 2.8 Data shown on computer printout of Accounts Payable Invoice Register.

indicated on the particular invoice to the applicable job. The job system number allows a further sort to apply the cost to a particular system or division of each job as established on the estimate. All this information then becomes part of job cost for cost accounting purposes.

The general ledger account code establishes the charge for general ledger purposes and will indicate either an inventory account, other asset account, or expense account. This information is then summarized by the computer and entered in the general ledger accounts through an interface program which automatically makes the entry to the proper account, or the summary may be used to make a separate manual entry to the proper accounts. The balances of the general ledger accounts are then used to prepare the trial balance and subsequent entries on the balance sheets and statements of profit and loss to establish the financial position of the company.

The invoice amount is the amount charged for that particular invoice to job cost (see illustration of job cost master, Fig. 2.9), entered in the general ledger, and accumulated to the YTD purchase amount on the vendor master.

The invoice terms and discount entry allows the computer to set up a payment due date which will allow the company to take advantage of the cash discounts allowable by the supplier or vendor.

The check register is a list of checks drawn by the computer to apply to the liability set up by the accounts payable amount on the invoice register. The check register will contain the following information:

1. Company check number
2. Vendor or supplier name and ID number
3. Vendor or supplier applicable invoice number
4. Gross amount of invoice
5. Discounts allowed
6. Discounts taken
7. Amount of check

A sample of a typical check register is shown as Fig. 2.10.

The company check number entry serves three purposes: controlling the checks issued by the company, verifying payment of invoices, and checking cleared checks against the company's bank statements for reconciliation purposes.

The vendor or supplier's name and invoice number identify the payment and allow the vendor or supplier to properly apply the payment on their records.

The gross, or total, amount of the invoice is the amount which will

Job Cost Master

1. Job prefix _____ 2. Job no. _______ 3. Job suffix _____

4. Job name ____________________

5. Contractor ____________ 8. PO/contract no. ______

6. Address ____________ 9. Estimator ______

7. City/ST/zip ____________ 10. Complete Y/N ______

11. Cost plus Y/N ___ 12. Job type ______ 13. Est. Cost ______

14. Contract amt. ____________________

	Estimate	Current	Last month
Labor hours			
Labor dollars			
Labor adder			
PO material			
Stock material			
Tool rental			
Subcontract			
Direct job expense			
Sales tax			
Total cost to date			
Total contract amount			
Percent complete			

Display Next Previous Enter Change Remove Help Quit

Figure 2.9 Job Cost Master.

be debited against the Accounts Payable account in the general ledger and used for balancing the various vendor or supplier accounts payable in the subsidiary accounts payable ledger, which lists all entries made to each vendor's account.

The discount allowed entry establishes the total amount of discount which is available if all accounts are paid according to the invoice terms.

The discount taken entry indicates the amount of discount which has been deducted from the total amount of the invoice being paid.

The computer will deduct the discount taken from the gross amount of the invoice and draw a check for the net amount. The discount taken amount will be applied to an income account in the general ledger and the net amount of the check will be deducted from the bank account in the general ledger. This amount will be used when reconciling the bank account to bank statements.

Check Register							
Check no.	Vendor code no.	Vendor name	Invoice no.	Invoice amt.	Dis-count allowed	Dis-count taken	Check amt.
6000	1001	ABC Supply	1001	$45,623.00	$456.23	$456.23	$45,166.77
6001	1125	B&J Supply	69801	$ 545.30	$ 10.90	$ 10.90	$ 534.40
6002	1746	Greg Supply	14207	$ 1,234.54	$ 24.69	$ 24.69	$ 1,209.85
6003	1803	Haps Mechanical	1004	$ 225.46	$ 2.25	$ 2.25	$ 223.21
6004	2130	J&J Electrical Sup	2968	$ 2,540.91	$ 50.81	$ 50.81	$ 2,490.10
6005	2645	Now Electric Suppl	24356	$ 1,567.32	$ 15.67	$ 15.67	$ 1,551.65
6006	2780	Owl Office Equip	18902				$ 0.00
			18945	$ 139.98	$ 1.40	$ 1.40	$ 138.58
6007	32654	Perfect Electric	32654	$ 5,678.00	$113.56	$113.56	$ 5,564.44
6008	1587	Walts Electric	1587	$ 254.95	$ 0.00	$ 0.00	$ 254.95
							$ 0.00
	Total			$57,809.46	$675.51	$675.51	$57,133.95

Figure 2.10 Check register.

Note that in the example in Fig. 2.10 the company's checks are listed on the check register in numerical order, which facilitates reconciliation of the company bank account with the bank's monthly statement. In addition the computer has sorted all due accounts payable invoices in alphabetical order by vendor and if more than one invoice is due to a particular vendor only one check has been drawn to cover the total for all the invoices.

Sales journal

The sales journal will contain the following information, which the computer will compile from the company invoices as they are drawn and listed on the invoice register:

1. Company invoice number
2. Customer name
3. Customer address
4. Job name
5. Date
6. Billing amount

Refer to Fig. 2.11 for an example of the sales journal.

The invoices are entered in the sales journal and will carry the following information:

XYZ Electric
Sales Journal

Month: January 1993

Invoice no.	Date	Customer name	Address	Job no.	Amount
98765	01/03/93	Some General	245 Any St.	30492	$49,182.00
98766	01/03/93	Smith Const.	41 S. 1st St.	21392	1,200.00
98767	01/04/93	Jones Paving	107 W. New Rd.	10393	35,000.00
98768	01/04/93	Bob Contract	235 N. 2nd St.	41092	2,500.00
Total billing					$87,882.00

Figure 2.11 Typical sales journal entries.

1. Company name and address
2. Company invoice number
3. Date
4. Terms of payment
5. Customer name and ID number
6. Customer address
7. Job name
8. Your company's job number
9. Location of job and customer's job number
10. Billing detail as required by the contract
11. Total billing to date
12. Previous billing
13. Net billing for this period
14. Previous billing not paid
15. Total amount now due and payable

Figure 2.12 shows an example of the company invoice.

The company name and address are necessary to identify your company to your customer. The company invoice number is also for identification purposes in establishing the accounts receivable records and also to establish an audit trail. The invoice number will also be used to locate the invoice in case of need to verify billing or to answer questions that may arise in the future.

The invoice date establishes the date of billing and the date terms of payment start. The date may also be used to determine the period ap-

INVOICE
XYZ ELECTRIC
1410 SOME ROAD
NEW TOWN, MI 19210
Phone (111) 234-5678

INVOICE NO. 96765

Date 01/03/93 Terms: Net 30 days

Customer no. 8096

Bill to: Some General Contractor
245 Any Street
Home Town, MI 19235

Job no. 123456

Job name: Home Town Airport Our job no.: 30493 Location: 23000 Lone Road

Contract amount $3,456,123.00

Completed to date, 30%	$1,036,836.00
Less: previous billing	987,654.00
Net billing this period	$ 49,182.00
Previous billing not paid	50,000.00
Total amount now due and payable	$ 99,182.00

Figure 2.12 Example of company invoice.

plicable to this billing, and is used to allow the computer to age the accounts receivable.

The terms of payment as noted on the invoice informs the customer of the dates by which payment must be received for the customer to be entitled to applicable discounts and not be in arrears.

The customer's name, code number, and address are required not only for mailing purposes but also to set up your company's accounts receivable records. The accounts receivable posting becomes an automatic operation each time billing invoices are entered.

The "job name" entry on the invoice identifies the job on which you are billing to the customer. Your company's job number is for the use of your company's personnel in identifying the job in your estimating and accounting records.

The job physical location and customer job number aid your customer in identifying the particular job for which you are billing.

Accounts receivable

Information required for completion of the Accounts Receivable Job Master (see Fig. 2.13) such as total (TTL) invoices, last billing date,

Accounts Receivable
Job Master

1. Job prefix ______ 2. Job no. ________ 3. Job suffix ________

4. Job name ____________________ 5. Contract no. ________

Bill to: ____________ Group ____________ GL account ________

____________ Job type ____________ Invoice type ________

____________ Sub/prime ____________ City ________

____________ Tax rate ____________ Tax method ________

Final bill Y/N Prog. Annually Flag Y/N

Base contract amt. ________ Change-Order no. ________ Change-Order amt. ________ Current contract amt. ________

Labor adder ________ TTL invoices ________ Last bill date __/__/__

Overhead ________ Current due ________ Last Pmt. Date __/__/__

Profit ________ Retention ________

Display Next Previous Enter Change Remove Help Quit

Figure 2.13 Accounts Receivable Job Master printout.

current amount due, retention amount, and last payment (pmt.) date will be applied and/or accrued as required. The Accounts Receivable Job Master then becomes the company's record of contract amount, Change-Order amount, and total amount of billing to date. All the other information on the Accounts Receivable Job Master will have been previously entered when the contract for the job was received.

As the invoices are being prepared, the entry of the job prefix and job number will cause the computer to access the Accounts Receivable Job Master for the job name, the contract number, and the customer's billing address and apply this information directly to the invoice. Also, the computer will access the job type, invoice type, subcontract or prime contract, tax rate, tax method, and progress analysis information shown on the Accounts Receivable Job Master and use this information to set up the correct invoice form for that particular job.

The labor adder, overhead, profit, and retention information shown on the Accounts Receivable Job Master will also be applied to the cost amounts shown on the invoices.

Depending on the contract requirements the detail required for billing is variable: any of, and in some cases all of, the following items may be required:

1. Separate items shown for labor and material charges
2. Direct job expense
3. Statement of overhead charge
4. Amount of profit being billed
5. Percent of completion being billed
6. Architect's certification of accepted work (required by many contracts)
7. Certified labor reports (required by some contracts, particularly those with government agencies)

In all cases it is advisable to ascertain that all conditions required by contract and applicable to your billing are met on each and every billing, as this will help to avoid many delays in receiving payment on your invoices.

Most projects or jobs of any size require several months to complete and are billed on a progressive basis as portions of the work are completed. The job cost is accumulated from the beginning of the job in the job cost records as a total to date. Thus, when billing, the first item to appear will naturally be total billing to date. Then to arrive at current period billing, it is necessary to subtract the total of previous billings to arrive at the net billing for the current period.

Each of the above will be shown as a separate item on the invoice. Then any previous billings which have not been paid will be shown and added to the current billing figure to arrive at the figure for the total amount now due.

The sales journal is the source of information for regular income entries in the general ledger, and as with the invoice register for accounts payable, the information from the sales journal may be posted automatically to the income section of the general ledger through an interface program. Alternatively, the sales figures may be picked up from the journal by a clerk and posted as a separate item to the general ledger accounts.

The accounts receivable entries to individual customers' job accounts should be made automatically from the sales journal by the computer as this will eliminate the opportunity for error which is always present when a clerk is required to transfer information from one journal to another. The invoices from the sales journal will be entered by date and invoice number in the correct dollar amount to each customer's account in accounts receivable.

The accounts receivable ledger is a subsidiary ledger to the general ledger. As stated above the debit entries to accounts receivable are originated in the sales journal. The accounts receivable ledger reflects money owed to the company by others and is one of the most valuable and important records of the company. A company's accounts receivable are one of its primary assets, and accounts receivable records should be protected from all sources of destruction or unauthorized alteration.

Credit entries to customers' accounts in the subsidiary accounts receivable ledger originate from the cash received register and occasionally from a company credit memo issued to correct an error in a customer's account.

At the end of the month when all debit entries and credit entries have been made to the accounts receivable, the computer will prepare a listing of all unpaid or open accounts and indicate those items which are current, and those which are 30, 60, and 90 days old. This aging list is then used according to company policy to begin efforts to collect the past due items.

Many companies begin this collection process with a firm but friendly letter to the customer requesting payment of the past due account. This letter is generally the first of a series of three form letters which can be stored in the computer and then printed and directed to the proper customer through a mail merge program. This eliminates the constant rewriting of collection letters by the clerical staff.

Payroll

Of all the functions which a computer is capable of performing for the business, the payroll function is the largest timesaver. The initial employee record to be set up in the payroll program will contain the following information:

1. Employee's name
2. Employee's company payroll number
3. Employee's address and social security number
4. Employee's pay rate (either hourly or periodic) and applicable tax information
5. Employee's status (hourly or salary)
6. Employee's job classification
7. Number of dependents and other personal information
8. Earned vacation classification
9. Sick leave classification

10. Other benefit information
11. Union benefit deductions and other deductions
12. Employee's date of hire
13. Any other items required

See Fig. 2.14 for a sample of an Employee Master.

Once the Employee Master is completed for all employees, only the following information needs to be entered for each employee for each week:

Figure 2.14 Employee master printout.

Employee Master

1. Employee no. ________ 2. Name ________

3. Address ________

4. City/ST/Zip ________

5. SSN ___ ___ ___ 6. Telephone ( ) ___ -______

Sex	EEOC	Dept.	Marital stat.	W/H Dep.	St. Tax	Ex-empt.	Fixed fed.	Driver
___	___	___	___	___	___	___	___	___

16. Pay code ________ 17. Hourly rate ________ 18. Salary ________

19. Union code ________ 20. Profit share Y/N ____ 21. Birth date __ __ __

22. Hire date ________ 23. Term date ________

24. Car allow. ________

25. Vac. code ________ 26. Vac. hrs. accr. ______ 27. Sick hrs. accr. ______

Deductions

Code	Description	Amount	Limit
___	___	___	___
___	___	___	___
___	___	___	___
___	___	___	___
___	___	___	___

Display Next Previous Enter Change Remove Help Quit

(Continued)

	Gross	Federal	OASDI	Medicare	State
MTD	______	______	______	______	______
QTD	______	______	______	______	______
YTD	______	______	______	______	______

	Regular	OT	DT	Other	
					_____ Net
Hours	______	_____	_____	_____	_____ Gross
Amount	______	_____	_____	_____	_____ Federal
					_____ OASDI
Check date __/__/__					_____ Medicare
Check no. ______					_____ State
Check type ______					_____ Ded. 1
Sick hrs. ______					_____ Ded. 2
Vacation hrs. ______					_____ Ded. 3
					_____ Ded. 4
					_____ Ded. 5

Display Next Previous Enter Change Remove Help Quit

Figure 2.14 (*Continued*)

1. Employee number
2. Number of hours worked
3. Job number applicable to hours worked
4. Labor codes for all hours worked
5. General ledger charge code

The computer entry screen for weekly payroll information is illustrated in Fig. 2.15. All the following management information reports are automatically furnished by the computer when the weekly payroll entries are made:

1. *Payroll register:* The Payroll Register and payroll checks will be automatically printed by the computer. The payroll check register becomes the company's permanent payroll record for audit purposes and balancing record for all other payroll reports. See Fig. 2.16 for a sample Payroll Register.

Employee no. ______ Job no. ______ Labor code ______ System no. ______

Div. no. _____ Regular hours _____ OT hours _____ DT hours _____ Other _____

Figure 2.15 Data appearing on computer entry screen for weekly payroll.

Payroll Register
XYZ Electric

Date 03/14/93

Check no.	Employee name	Emp. no.	Time: Regular	Time: OT	Time: DT	Other	Gross	OASDI/ Medicare	Federal	State	Dues	Deducts	Net
1000	J. Arbor	1006	40.0	0	0		$ 560.00	39.76	112.00	11.20	11.20		385.84
1001	F. Bach	1201	29.0				406.00	28.82	81.20	8.12	8.12		279.74
1003	P. Cash	1304	40.0				560.00	39.76	112.00	11.20	5.00		392.04
Total			109				$1,526.00	$108.34	$305.20	$30.52	$24.32		1057.62

Figure 2.16 Computerized payroll check register.

2. *State and federal tax withholding:* The state and federal tax withholding amounts are available from the payroll check register each pay period to allow the required deposits to be made to the taxing agency. The periodic amounts will be accumulated by the computer and then used with the necessary detail to prepare the applicable reports, such as the federal 941 withholding tax return. The preparation of these reports will be handled automatically by the computer.

3. *FICA withholding:* Federal FICA (Federal Insurance Contribution Act) reports are prepared by the computer in the same manner as the withholding tax returns, with the same schedules by employee name being used for both reports.

4. *Federal unemployment tax information:* Calculations required for FUTA (federal unemployment Tax Act) returns are taken from the federal 941 reports.

5. *State unemployment tax information:* The SUTA (State Unemployment Tax Act) returns, which require each employee's name and earnings, are accumulated and prepared by the computer without further entry.

6. *Union and benefit information:* Union or other organizational reports such as union dues, union pension, industry pensions, and other benefit reports are prepared from the information in the Employee Master and the periodic payroll registers. All these are an automatic function of the computer.

7. *YTD wages and accumulated leave:* Employee MTD, QTD, YTD gross wages and tax information, and accrued vacation and sick leave time, are entered to the Employee Master automatically by the computer.

8. *Job Cost Master:* Job cost labor reports are accumulated and printed using information from the single payroll entry made each period for each employee. The job cost labor reports will show each employee working on all applicable jobs by employee number, job number, number of hours worked, amount paid, and the proper labor code. All the labor information for job cost is automatically posted to the Job Cost Master.

All the reports and information noted above are accurate and do not require any hours of checking time by the staff. Once the payroll check register has been balanced, all the resulting reports are correct and require no further audit.

Cash received journal

All funds received by the company should be recorded in the cash received journal daily as received. The information recorded should include the payer's name and code number, applicable job number and/or general ledger account number, and for bank deposit information the issuing bank's ID number.

The deposit slip, which will be printed by the computer for your bank, will of course show only the issuing bank's ID number and the amount being deposited for each check. Cash deposits will be individually identified in the cash received journal with the same information as shown on check deposits.

The computer will automatically credit payments received from your customers directly to their accounts in your accounts receivable subsidiary ledger records using the customer code number and the amount recorded to each job.

The amounts shown as deposits in the cash received journal are used to verify the bank's record of your deposits as shown on the bank statement each month.

Reconciliation of the bank account becomes a relatively simple matter when you have the deposit slips generated by the cash received journal to verify all deposits and the check register against which you can check off all checks that have cleared the bank. Figure 2.17 shows a sample input screen for the cash received journal.

Fixed asset schedule

The recurring calculation-of-depreciation nightmare experienced by the accounting departments of most businesses has been solved by the inexpensive fixed assets programs now available for the PC. All these programs allow you to make one entry for each fixed asset, noting the asset class, the cost, depreciable amount, and type of depreciation applicable to that particular asset, and accrued depreciation to date.

From that point on a simple update procedure is performed each period and the computer collates current depreciation for all assets listed and also lists total depreciation to date by asset. The computer print-

Date __/__/__ Page 1 of __

Name	Job no.	Job name	Date paid	Inv. no.	Amt.	Routing slip	GL Acct.	Check total
___	___	___	___	___	___	___	___	___
___	___	___	___	___	___	___	___	___

Figure 2.17 Input screen for cash received journal.

out is then used to back up general ledger depreciation amounts both for book and tax purposes.

All the programs have the ability to sort and list assets by class, which produces a complete, detailed listing of each class of assets. Figure 2.18 illustrates a typical schedule of fixed assets and depreciation expense report which is created automatically by the computer at the end of any accounting period. It should be noted that the computer is capable of using this information to print out a simple list of the fixed assets showing only the asset number, description, date acquired, and the acquisition value.

General ledger

The computerized general ledger is set up in the same manner as in a manual system and divided according to the four basic categories: assets, liabilities, income, and expense. It is necessary to have a chart of accounts, which of course is expandable, and a series of logical account codes. One primary difference in the account codes between the manual system and the computerized system is that the computer system requires a decimal series of code numbers, while the manual mode allows an alphanumeric system. For example, the fixed-asset account in the computer system would appear as:

180.0 Buildings

180.1 Allowance for Depreciation—Buildings

In the manual system the same accounts might appear as:

180 Buildings

180R Reserve for Depreciation—Buildings

One reason for this difference in account coding is that it is much easier to program the computer to recognize numerical fields than it is to program it to read and process alphanumeric information. Another is easier input from a 10-key device. The input screen for the general ledger is shown in Fig. 2.19.

General ledger asset accounts. The information contained in the asset accounts will originate from the various journals described earlier in this chapter. For example the Cash in Bank account will receive debit entries automatically from the cash received journal. Credit entries to the Cash in Bank account will be posted directly from the check register. The asset accounts are as shown in Fig. 2.20.

The general ledger Accounts Receivable account will receive debit entries from the sales journal posted in total amounts. Credit entries

Schedule of Fixed Assets

Date 3/01/91

Asset class	Asset no.	Asset description	Date acquired	Acquisition value	Depr. method	Life, yr mo	B/salvage L sect. 179	Depreciable basis	Last depr.	Prior total accum.	Depr. this run	Current YTD depr.	Total accum. depr.
1	1200	Building #1	3/8/88	$75,000	SL	30		$75,000.00	$2,500.00	$5,000.00	$2,500.00	$2,500.00	7,500.00
3	234	Chev P/U	3/1/90	12,000	DDB	4	B	10,000.00			5,000.00	5,000.00	5,000.00
4	555	Bender	1/1/91	6,000	SL	3	B	6,000.00			334.00	334.00	334.00
2	2415	Desk	3/1/88	250	SL	10		250.00	50.00	50.00	25.00	25.00	75.00
Total				$93,250.00				$93,250.00	$2,550.00	$5,050.00	$7,859.00	$7,859.00	$12,909.00

Figure 2.18 Typical schedule of fixed assets and depreciation expense report created automatically by the computer at the end of any accounting period.

Journal entry no.	Date	De-scription	Acct. no.	Acct. title	Debit	Credit
			Account distribution			
1	3/01/92	A/P	201	A/P	3,000.00	
			101	Cash in Bank		3,000.00
2	3/01/92	A/R	101	Cash in Bank	2,500.00	
			110	A/R		2,500.00
3	3/01/92	Sales	110	A/R	15,250.00	
			302	Income		15,250.00
And so forth.						

Figure 2.19 Input screen for general ledger.

General Ledger
Asset Accounts

Acct no.	Acct. name
	CURRENT ASSETS
100	Cash on Hand
101	Cash in Bank
102	Certificates of Deposit
105	Accounts Receivable, Regular
110	Accounts Receivable, Progress
111	Accounts Receivable, Retail Sales
115	Notes Receivable, Current
120	Inventory Stock
120.1	Inventory Progress
140	Unbilled Amounts on Completed Contracts
140.1	Costs and Estimated Earnings on Contracts in Progress in Excess of Billings
150	Prepaid Expenses and Other Current Assets
	OTHER ASSETS
151	Land Held for Investment
152	Cash Surrender Value of Life Insurance
153	Investments (Stocks and Bonds)
	FIXED ASSETS
160	Land
161	Buildings and Improvements
162	Office Equipment
163	Furniture and Fixtures
164	Autos and Trucks
165	Construction Equipment
166	Tools

Figure 2.20 Asset accounts.

to the Accounts Receivable account will be posted directly from the total of the cash received journal applicable to accounts receivable.

The general ledger Fixed Asset account will generally receive entries from the Accounts Payable Invoice Register. These entries will reflect the purchase of vehicles, office furniture and equipment, shop equipment, and so forth. Purchases of land and/or buildings are usually entered from an entry in a special general journal.

Other assets such as notes receivable and amounts due from officers are also usually handled through a special general journal entry.

General ledger liability accounts. Accounts Payable is the primary liability account and receives credit entries from the Accounts Payable Invoice Register. Debit entries to this account are posted direct from the cash disbursed or check register. As noted above the entries to the individual vendor accounts are made by the computer at the same time as it is accumulating the total for posting to the General Ledger Account.

Other liability accounts such as Notes Payable and other borrowing accounts are generally set up through special general journal entries. The general ledger liability accounts are as shown in Fig. 2.21.

General ledger income accounts. Income from sales or operations is generally the predominate source of income. The information for this account will be derived directly from the sales journal and may be entered to the general ledger each time sales invoices are posted to the invoice register. This entry may be from an interface in the program or manually.

Entries to the Interest Income account most often are made as a special entry from the general journal. Entries to Discounts Earned are derived from the check register. Other Miscellaneous Income is picked up from the general journal. See Fig. 2.22.

General ledger expense accounts. All the information for the expense accounts comes directly from the Accounts Payable Invoice Register. This information can be automatically summarized by account code by the computer and through an interface be posted directly to the various expense codes. It is possible to manually summarize the information and post it to the appropriate accounts manually. The normal expense accounts are listed in Fig. 2.23.

Financial reports

Balance sheet. Once all the entries for the period have been made to the general ledger the computer will sort, adjust, and accumulate each account to the current balance.

Then a trial balance is obtained to ascertain that the general ledger

General Ledger
Liability Accounts

Acct. no.	Acct. name
	CURRENT LIABILITIES
205	Accounts Payable
206	Retention
208	Accrued Payroll
209	Payroll Taxes
210	Union Benefits
211	Amounts Withheld from Employees
215	Additional Costs on Completed Contracts
216	Amounts Billed in Excess of Costs and Estimated Earnings on Contracts in Progress
218	Current Income Taxes
219	Deferred Income Taxes Owed
220	Current Portion of Long-Term Debt
221	Other Taxes Due, Current Period
	OTHER LIABILITIES
225	Mortgages
226	Notes Payable Later Than 1 Year
	STOCKHOLDERS EQUITY AND NET WORTH
260	Common Stock
261	Additional Paid-In Capital
262	Retained Earnings

Figure 2.21 Liability accounts.

General Ledger
Income Accounts

Acct. no.	Acct. name
	INCOME
301	Income, Service Jobs
302	Income, Progress Jobs
303	Income, Retail Sales
310	Interest Income
315	Discounts Earned
329	Income, Disposition of Assets
340	Other Miscellaneous Income

Figure 2.22 Income accounts.

General Ledger Expense Accounts	
Acct. no.	Acct. name
	EXPENSE
501	Advertising and Promotion
502	Automobiles and Trucks
503	Bad Debts
504	Charitable Contributions
505	Collection Expense
506	Depreciation and Amortization
507	Dues and Subscriptions
508	Education Expenses
509	Overhead Employee Benefits
510	Freight and Express
511	Heat, Light, Power, and Water
512	Company Insurance—General
514	Overhead, Employee Workmens Compensation
515	Public and Employers Liability Insurance
516	Legal and Accounting
518	Miscellaneous
520	Office Supplies and Expense
521	Pension and Profit-Sharing Plans
522	Rent
523	Repairs and Maintenance
525	Plan, Bid Bonds, Estimating, and Engineering Expense
526	Salaries and Wages—Overhead Personnel
528	Shop Supplies and Expense
530	Small Tools
532	Taxes and Licenses—General
534	Taxes—Overhead Payroll
536	Telephone, Telegraph, and Postage
537	Travel—Administrative
550	Other Expenses

Figure 2.23 Expense accounts.

is in balance. In most cases the computer will print out the trial balance in a matter of 30 minutes or less. At this point any adjustments such as current earnings or loss should be entered on the proper accounts in the general ledger through the general journal.

The current period balance sheet and statement of profit and loss will then be assembled and printed automatically by the computer. A sample balance sheet and statement of profit and loss are shown in Figs. 2.24 and 2.25.

Purchasing

A tightly controlled and efficient system of purchasing is vital to the successful operation of any business, but this is especially true in the elec-

Figure 2.24 Typical balance sheet.

ABC Electric
Balance Sheet
For Period Any Month End
Assets

Current Assets	
Cash	$ 50,000
Certificates of deposit	165,000
Accounts receivable	75,000
Unbilled amounts on completed contracts	10,000
Material and supplies inventory	20,000
Costs and estimated earnings on contracts in progress in excess of billings	40,000
Prepaid expense and other current assets	5,000
Total current assets	$365,000
Other Assets	
Land held for investment	$ 75,000
Cash surrender value of life insurance	16,814
Total other assets	$ 91,814
Fixed Assets	
Land	15,000
Building and improvements	65,000
Construction equipment	75,000
Vehicles	3,000
Office equipment	4,000
Total fixed assets	$162,000
Less: Accumulated depreciation	60,000
	102,000
Total assets	$558,814

(Continued)

trical contracting business. The historically small margins the electrical contractor is able to obtain due to the bidding process compels him to purchase every item to his best advantage and at the proper time to control material delivery to the job and to control cash flow.

The reports available from the original estimate will establish the time frame for purchase of the materials for any particular job. The estimate supplies a consolidated materials list (a Bill of Material) for use in obtaining vendor bids on the required material. A comparison of supplier or vendor bids on the Bill of Material allows those responsible for purchasing to determine the best available overall price.

As each purchasing decision is made a Purchase Order should be issued and entered in the computer on the Purchase Order Register. The Purchase Order Register should contain the following information:

ABC Electric
Balance Sheet

Liabilities	
Current Liabilities	
Accounts payable, including retention of $10,000	$ 95,000
Accrued payroll, payroll taxes, and benefits	2,000
Addition costs on completed contracts	2,500
Amounts billed in excess of costs and estimated earnings on contracts in progress	54,500
Other current liabilities	5,000
Income taxes	
Current	4,500
Deferred	10,000
Current portion of long-term debt	6,500
Total current liabilities	$180,000
Long-Term Debt	$ 80,000
Stockholders Equity—Net worth	
Common stock, par value $10.00 per share, authorized 10,000 shares, issued and outstanding 2,000 shares	$ 10,000
Additional paid-in capital	20,000
Retained earnings	268,814
	$298,814
Total liabilities and net worth	$558,814

Figure 2.24 *(Continued)*

1. Purchase Order number
2. Date Purchase Order issued
3. Supplier's name
4. Supplier's address
5. Applicable job number

Each Purchase Order issued should contain at least the following information:

1. Company name and address
2. Delivery address
3. Job name
4. Job number
5. Supplier's name and address
6. Credit terms

ABC Electric
Statement of Profit and Loss
For Period Ending __/__/__

			Percent of sales
Sales		$2,600,000	100.00
Cost of sales:			
Material	$936,000		36.00
Labor	728,000		28.00
Payroll tax	214,240		8.24
Direct job expense	104,000		4.00
Subcontract	91,000		3.50
Prime cost		$2,073,240	79.70
Gross profit		526,760	20.26
Overhead costs		473,200	18.20
Operating profit		$ 53,560	2.06
Income taxes		18,746	.72
Net profit		$ 43,814	1.34
Retained earnings, beginning of year		$ 225,000	
Retained earnings to date		$ 268,814	

Figure 2.25 Typical statement of profit and loss.

7. Cash discount amount or percentage
8. Required delivery date
9. Quantity ordered
10. Detailed description of material
11. Unit price
12. Extended price
13. Special instructions if any
14. Signature of purchaser
15. Company purchasing policy on reverse side of order

A Purchase Order becomes an enforceable contract when issued to a supplier, so it is extremely important that all the information noted above be on the order. The presence of this information will be required to settle any dispute that may arise concerning quality of goods, pricing, delivery dates, and discounts allowed. Figure 2.26 is a typical Purchase Order for an electrical contractor.

ABC ELECTRIC
404 Any Street
New Town, MI 40801

(302) 234-4567

PURCHASE ORDER

Date __ / __ / __

To:

____

____

____

____

____

Deliver to:

Name ______________________

____

Street ______________________

City ______________________

State ______________________

Phone (____) ____-________

Job name __________ Job no. ____________ Required delivery date__/__/__

Cash discount, % ____ Terms ________ FOB____________

Quantity	Description	Unit	Unit cost	Price

Special instructions ______________________________

____

Signature ______________________

Title ______________________

Figure 2.26 Example of Purchase Order.

An expediting record should be maintained for each job, showing:

1. Purchase Order number
2. Date issued
3. Job name and number
4. To whom issued
5. Brief description of material ordered
6. Date required
7. Space for dates follow up made
8. Space for comments on follow up
9. Date material received

This expediting record, which is a form entered in the computer, remains active in the computer until the job is completed and all Purchase Orders closed out. A hard copy of the form can be printed at any time. This form will also serve as the purchasing department's receiving record for each Purchase Order. Figure 2.27 represents a Purchasing and Expediting Record.

A receiving record signed by the company employee who actually received the material from the delivering carrier will be forwarded to the accounts payable department. The receiving document is to be matched with and recorded on the company Purchase Order. The receiving document is then attached to the vendor's invoice to form a package eligible for payment.

Tool Control

The ability of the electrical contractor to supply his job with the proper tool at the time needed is of equal importance with having the

PURCHASING AND EXPEDITING RECORD

Job name ______________ Job no. _______ Proj. manager ___________

PO no.	PO date	Date required	Date promised	Contact date	Contact date	Date received	Comments
1234	1/5/92	3/10/92	3/10/92	1/20/92	2/10/92	3/09/92	
1345	1/9/92	4/10/92	4/10/92	1/23/92	2/25/92		will ship 4/10
1544	2/1/92	5/5/92	5/5/92	2/25/92	3/25/92	5/5/92	

And so forth.

Figure 2.27 Purchasing and Expediting Record.

correct material on the job when required. The complexity of present day electrical installation on all jobs is such that many tools must be furnished by the electrical contractor.

In many instances one of the first requirements for the electrical contractor will be to furnish temporary power and lighting for the project. Due to federal regulations this is no longer just a case of furnishing a few extension cords to be scattered around the job with some lamp sockets hung here and there. Certain lighting levels are now specified and must be met for safety reasons. In addition, sophisticated power panels with ground-fault protection must be supplied. These panels range in cost from $1,500 to $3,000 each, and a large project may require 100 or more of them. This is just one example of the large tool investment required of the successful electrical contractor.

As is true with any investment, the successful company will have a complete record. In the case of tools this is accomplished by creating a tool register which lists all the tools which the contractor has available. Each tool should be assigned a number which identifies it by class and as an individual tool within the class. This may be a number system such as:

Class	code	Description	Date acquired	Cost
1	1000-23	3/4-in EMT bender no. 23	1/10/89	$ 250
2	1001-10	1-in EMT bender no. 10	5/06/90	$ 475
2	1020-05	2-in GRC bender no. 05	10/07/87	$ 2,500
6	3550-02	Trencher no. 02	1/15/90	$25,000

The listing continues until all tools have been assigned a number and entered in the computer. A tool control register listing all the tools is then printed and becomes the tool inventory. The computer input screen, Tool Master, is illustrated in Fig. 2.28, and the Tool Control Register itself would appear as shown in Fig. 2.29. It should be noted that the computer is capable of sorting the tools by class and listing them in tool number order by class for effective control of all the tools in the contractor's inventory.

Many large projects are now being constructed by what is termed the "fast track" method. Since these projects are designed as they progress, it is not possible to have a complete set of plans to examine and determine a fixed bid price. These projects require the contractor to work from preliminary drawings and propose a "not to exceed" price, from which construction will proceed on a cost-plus basis not to exceed a given figure.

The electrical contractor will encounter a very real job cost to supply the tools necessary to construct the job. In most instances the contractor will be allowed to furnish the smaller, less expensive tools re-

Tool Master

1. Group ________ 2. Suffix ________ 3. Company tool no. ________

4. Description ________________ 5. Serial ____________

6. Model no. ____________ 7. Quantity ____________

8. Status (new/used) ________________ 9. Status (date) ________________

10. Purchase (date) ____________ Purchase price ____________

11. Replacement cost ________________

12. Rental rates

Daily ________ Weekly ________ Monthly ________

Figure 2.28 Entry screen for Tool Master, used for tool control.

quired at a percentage of the estimated cost. The upper limit for the price of a "small" tool will vary by job, but may be in the range of $300 to $500. Tools costing more than the set amount will be charged to the job on a tool rental basis. Most cost-plus–type contracts will have a clause stating that tool rental rates must be in line with the prevailing rates in the local area. Each contractor must by law determine his own rental rates. These rates should be established according to his own formula.

The formula used to establish the contractor's rental rates should include such items as:

1. Tool replacement cost
2. Insurance
3. Depreciation
4. Handling expense
5. Storage expense
6. Transportation expense
7. Tool maintenance expense
8. Profit

After the rental formula has been established, the Tool Control Register should be used to identify all the contractor's tools, and a Tool Rental Schedule and Tool Catalog developed.

The Tool Rental Schedule will be printed by the computer using the information shown on the Tool Control Register (Fig. 2.29). The com-

Tool Control Register

Class	Company tool no.	Description	Mfg. serial no.	Model no.	Quantity	Purchase price, $	Purchase date	Replacement cost, $	Rental Rates $ Daily	Weekly	Monthly
1	1000-23	3/4" EMT Bender	12.90099	306	6	250.00	10/09/90	260.00	2.00	6.00	18.00
1	1001-10	1" EMT Bender	23456	602	2	475.00	9/10/89	575.00	5.00	15.00	40.00
2	1020-05	2" GRC Bender	526789	1550	1	2,500.00	1/20/92	3,000.00	20.00	50.00	150.00
6	3550-2	Trencher	98765	32	1	25,000.00	8/10/89	30,000.00	100.00	300.00	900.00

And so forth.

Figure 2.29 Tool Control Register.

puter will extract the following information and print a list showing for each tool only:

Tool no.	Description	Cost	Rent		
			Day	Week	Month

The Tool Rental Schedule may be used to negotiate and establish the contractor's tool rental rates before the contract is signed. This avoids future disputes as the contract billing progresses. In most instances it is possible to have the contract state that tool rental will be charged according to the electrical contractor's established rental rates. By using this method the contractor is assured of receiving enough revenue from his tools to allow him to replace them when needed and earn a reasonable profit.

A Tool Catalog containing the rental information plus a photograph of each tool is a powerful sales aid in negotiating your rental rates. In addition, listing the replacement cost of each tool in the catalog helps the contractor's employees gain respect for the value of the tools they are using.

The Tool Control Register and the Tool Rental Schedule are stored in the computer. The tool department maintains a simple In and Out Form on which they list each tool sent to a job as follows:

Job no.	Tool no.	Date out	Date returned
P12345	1002	1/25/92	When returned
P43216	1009	1/27/92	2/23/92
And so forth.			

The primary purpose of this form is to inform the billing department which tools are sent to each job, the date sent out, and the date the tools are returned.

The secondary reason for maintaining the In and Out Form is to maintain a record of the location of each tool at any given time. This, in turn, aids in the annual tool inventory by giving the tool department a list of tools which should be on each job.

The information on the In and Out Form enables the billing department to charge tools to applicable jobs by simply entering into the computer the job number, the tool number, and the date sent to the job. The computer will then compute the applicable rental to be charged to each job for all tools on the job. By entering the job number, tool number, and date returned, the computer will discontinue billing on the returned tools. The total billing to date on returned tools will continue to appear.

The tool rental billing generated by the computer would appear as follows:

Tool Rental Billing Date ___________

Period ending _________

Job no. Tool no. Description Date Date Prev
To date Current Out In Billed Billed Billing

This schedule will become a part of the regular monthly billing and be supplied to the customer for detail of the tool rental billing listed on the invoice.

Management Use of Computer Reports

The reports generated by the computer give management a level of control never before possible. Each week a list of jobs in progress and the names of the men or women assigned to each job will be compiled by the computer and furnished to management. This report allows management with the help of the general foreman on the job site to determine if the crew size should be increased remain the same, or be decreased, and if it is to be decreased which men should be transferred or discharged. The Weekly Labor List is shown in Figure 2.30.

Each month a comprehensive schedule of work in progress will be furnished to management. This schedule is generally referred to as the "Progress Analysis" (See Fig. 2.31 for sample) and will contain the following information:

Weekly Labor List

Job no.	Name	Job no.	Name
30189	J. Bill	4492	K. Archer
	K. Jones		
	F. Johnson	4692	D. Die
			L. Turner
30390	L. Smith		W. Zee
	R. White		A. Yee
31290	G. Iron		
	E. Jones		
	Y. Kenny		
	A. Long		

Figure 2.30 Weekly Labor List.

Figure 2.31 Progress Analysis.

Progress Analysis

Month of ______________________

	Man-hours	Labor	Payroll tax	Material	Direct Job Expense	Tools	Total
Job no. 10390							
Estimated	2,000	28,000.00	9,000.00	36,000.00	4,000.00	1,000.00	69,000.00
Used to date	500	7,000.00	2,240.00	10,000.00	1,500.00	200.00	18,700.00
% used to date	25%	25%	25%	28%	37.5%	20%	27.1%
Over (under) est.	(1,500)	(21,000.00)	6,760.00	(26,000.00)	(2,500.00)	(800.00)	(50,300)
Job no. 23456							
Estimated	5,000	70,000.00	22,400.00	95,900.00	13,000.00	10,300.00	211,600.00
Used to date	2,500	36,250.00	11,600.00	43,155.00	6,500.00	4,000.00	101,505.00
% used to date	50.0%	51.8%	51.8%	45.0%	50.0%	38.8%	47.9%
Over (under) est.	(2,500)	(33,750.00)	(10,800.00)	(52,745.00)	(6,500.00)	(6,300.00)	(110,095.00)
Job no. ______							
Estimated	______	______	______	______	______	______	______
Used to date	______	______	______	______	______	______	______
% used to date	______	______	______	______	______	______	______
Over (under) est.	______	______	______	______	______	______	______
Totals:							
Estimated	7,000	98,000.00	31,400.00	131,900.00	17,000.00	11,300.00	280,600.00
Used to date	3,000	43,250.00	13,840.00	53,155.00	8,000.00	4,200.00	120,205.00
% used to date	42.9%	44.1%	44.1%	40.3%	47.1%	37.2%	42.0%
Over (under) est.	(4,000)	(54,750.00)	(17,560.00)	(78,745.00)	(9,000.00)	(7,100.00)	(160,395.00)
Column no.	1	2	3	4	5	6	7

(Continued)

Progress Analysis

Month of

	Contract amt.	Total cost/ est. cost = % of comp.	Proj. mgr. est. % of comp.	Previous billing	Current % of comp. billing	Actual current billing	Total billing to date	Billing over (under) cost & profit
Job no. 10390								
Estimated	100,000.00							
Used to date		27%	30%	10,000.00	27,000.00	20,000.00	30,000.00	3,300.00
% used to date					27%		30%	
Over (under) est.								
Job no. 23456								
Estimated	259,300.00							
Used to date		48%	45%	38,895.00	124,464.00	90,000.00	128,895.00	4,431.00
% used to date					48%			
Over (under) est.								
Job no. ______								
Estimated								
Used to date	______	______	______	______	______	______	______	______
% used to date								
Over (under) est.								
Totals:								
Estimated	359,300.00							
Used to date		42.8%		48,895.00	151,464.00	110,000.00	158,895.00	7,731.00
% used to date					42.2%		44.2%	
Over (under) est.								
Column no.	8	9	10	11	12	13	14	15

Figure 2.31 (*Continued*)

1. Job number
2. Job name
3. Total estimated man-hours
4. Total man-hours used to date
5. Percent of estimated man-hours used to date
6. Total estimated labor cost
7. Total labor cost to date
8. Percent of estimated labor dollars used to date
9. Total estimated payroll taxes
10. Total payroll taxes accrued to date
11. Percent of estimated payroll taxes used to date
12. Total estimated material cost
13. Total material cost to date
14. Percent of estimated material used to date
15. Total estimated direct job cost
16. Total direct job cost to date
17. Percent of direct cost used to date
18. Total estimated tool cost
19. Total tool cost to date
20. Percent of estimated tool cost used to date
21. Total estimated cost
22. Total cost to date
23. Percent completion based on cost
24. Project manager's estimate of percent completion
25. Previous billing
26. Allowable current billing based on cost percent of completion
27. Actual current billing
28. Total billing to date
29. Over- and underbilling

By comparing items 23 and 24 on each job as illustrated in Fig. 2.31, columns 9 and 10, the manager has an accurate picture of job progress. If there is a variance of any consequence in either of these, the manager should compare the percentages in items 5, 8, 11, 14, 17, and 20 with items 23 and 24 to determine which elements of cost are running above estimate. This allows management and front-line supervision to investigate and take the steps necessary to correct cost overruns before they become a serious problem. Note on Fig. 2.31 that Job 23456 is showing labor hours used of 50 percent and labor used of 51.8 percent but that the percent of completion shown in column 9 is only 48 percent. The other elements of job cost, with the small exception of direct job expense, are in line with the project manager's estimate of percent completion.

Word Processing

Correspondence

Word processing always becomes one of the most heavily used applications in the computerized company. Shortly after beginning to use computer word processing most people will wonder how they ever managed with just a typewriter. Editing of a letter or any other document never requires retyping the entire page or document just to correct a word or even to insert new paragraphs.

Form letters become personally addressed letters, and the name and address are inserted automatically on each individual letter by using a merge list of addresses.

Forms

In addition, standard forms used by the company such as Change-Order forms, subcontract agreements, and requests for proposals become whole new documents by changing a few words or lines to fit the information to the particular situation.

All the popular word-processing programs now contain a spell checker that assures the user that they have no misspelled words in their document.

There are many word-processing programs on the market and most of them are very good. Word Perfect is probably the most popular. We recommend that you purchase a word-processing program rather than having your programmer attempt to write your own.

Desktop publishing

The programs available to accomplish desktop publishing, at the present time, all require that your computer and printer have the ability to compose and print graphics. A desktop publishing program will enable you to compose your own advertising copy and graphics for newspaper or flyer use, and this will save you considerable amounts of money over time, since you will not have to hire writers and artists.

Computer Equipment

Currently, the most efficient critical path method (CPM) system readily available is the 386, 20 megahertz, with a 170-megabyte hard drive, a 5¼-in floppy disk drive, a 3½-in drive, and a color monitor. The unit should have a graphics card to enable you to use desktop publishing and also to create graphs of your various accounting journals. As your company grows you will find it necessary to add additional units to handle the workload efficiently. Ultimately a network system which ties the whole computer system into an integrated whole will be the best solution.

The basic 386 CPM system will retail for about 550 hours of minimum wages. A suitable printer will be required, and for the smaller contractor a 24-pin printer is recommended. A good 24-pin printer will give you very close to letter quality printing and is also very fast while being economical. Laser printers are available at additional cost and are made primarily for high-volume printing requirements.

A few words of caution on equipment. The computer manufacturers continue to upgrade their equipment and the computer field advances very quickly, so quickly in fact that any equipment you purchase will be more or less outdated shortly after you install it. However, this does not mean that you need to be constantly upgrading your equipment. As long as the equipment you have is performing the function it was designed to do and is not obsolete and you can get the proper hardware or software support there is no need to replace it just because there happens to be a bigger and faster piece of equipment available. It is very expensive to implement, debug, purchase new forms, and retrain your personnel on the new hardware and software.

Funding

Use of the computer by a new business

One of the necessary steps in forming a new business is to obtain adequate financing to cover start-up costs and sustain the business until it begins to produce a profit. This can become a frightening process, especially if you need a loan, since the paperwork required often becomes a challenge to the new businessman.

The computer program called "First Step Review," developed by the Dallas-based National Business Association, can be of great help in solving this problem. First Step Review is a user-friendly program for use on IBM, IBM-compatible, and Macintosh computers. It outlines the SBA-backed loan program and helps people determine their chances for receiving such a loan. Much of the same information is required for your banker to allow the bank to grant a regular commercial loan.

The program requests information in five critical business areas from the user, then evaluates and scores the responses. A score of 70 or more suggests that the applicant has a good but not guaranteed chance of receiving an SBA or regular bank commercial loan.

The program is free and can be obtained by calling the association toll-free at 1 (800) 456-0440.

Two other programs, Cash Flow Analysis and Projected Profit or Loss Statement, are available from the association for $5 each. These two programs are helpful in developing these two components required in the loan qualification process.

Chapter

3

Effective Financial Management

It is an undeniable fact that operating a business must result in a profit being made or the business will fail. *Profit* is the amount remaining after all costs of doing business are paid. The very nature of the electrical contracting business is such that at no point during the active life of the business can the contractor reach a stage where all jobs are complete, all bills paid, and all fixed assets disposed of and what is left is his profit. Because of this, it is imperative that either the contractor or someone on his staff be able to read and thoroughly understand a balance sheet and statement of profit and loss. In addition the contractor must be able to utilize various financial ratios to the advantage of the business.

Financial Methods

A profit will be realized only when several factors have worked successfully together in the operation of the business. Some of these are:

1. Having a sufficient volume of sales to produce income which covers not only costs but overhead, and also leaves a margin of profit. This requires good estimating and sales techniques.
2. Having established job management methods which consistently enable the company to better or at least equal the estimated costs.
3. Having an effective implementation of financial controls which are applied on a consistent basis.

Proper financial management methods require that the following steps be taken:

1. Obtain maximum profitable sales volume within the limits of the working capital of the company.
2. Increase working capital by exploring all reasonable sources of supplementary financing to enable the business to properly finance an ongoing, increasing volume of sales while providing the services that are necessary to remain in a competitive position in the industry.
3. Control overhead costs so that they are on a par with or less than those of the contractor's competitors. Not only must the overhead be controlled but the contractor must ascertain that these costs are properly applied to all estimates.
4. Constantly monitor the financial standing of the business and compare it with other businesses in the market area as to the *five essential cost elements* of the business: material cost, labor cost, direct job expense, overhead, and profit.
5. Closely control the cost of inventories, vehicles, tools, and other equipment.

It has been established by Dun and Bradstreet after many studies that the principal cause of business failure is poor management and inadequate finances. Proper financial management results in proper amounts of capital being applied to working capital, fixed assets, and inventory. A stable financial condition can be achieved only when it is possible to acquire working capital in amounts that allow the company to perform a volume of services which produces a reasonable profit and return on investment.

The successful financial management of a business is contingent on many factors such as having adequate sales volume to ensure a profit, succeeding in regular and reasonably prompt collection of accounts receivables, and establishing strict methods of controlling and balancing the ratios and percentages established as norms in the descriptions which follow.

No matter what the volume of sales and the booked profit thereon, working capital tied up in old receivables becomes a cost factor and cannot be used to increase the income of the business.

Working Capital

In accounting terms, *working capital* is represented by the positive difference between the sum of current assets and current liabilities. Assuming that receivables are collected in a timely matter it is generally accepted, in the electrical contracting industry, that the comfortable ratio of sales volume to working capital is 6:1 to 8:1, with an upper limit of 10:1.

Current assets consist of the following accounts on the balance sheet:

1. Cash on Hand
2. Cash in Bank (including certificates of deposit)
3. Accounts Receivable (including retention)
4. Notes Receivable (due within 1 year)
5. Unbilled Amounts on Completed Contracts
6. Material and Supplies Inventory (warehouse inventory)
7. Costs and Estimated Earnings on Contracts in Progress in Excess of Billings
8. Prepaid Expenses and Other Current Assets

Current liabilities are:

1. Accounts Payable (including retention if any)
2. Payroll accounts: Accrued Payroll, Payroll Taxes, Union Benefits; Amounts Withheld from Employees
3. Additional Costs on Completed Contracts
4. Amounts Billed in Excess of Costs and Estimated Earnings on Contracts in Progress
5. Current Income Taxes
6. Deferred Income Taxes Owed
7. Current Portion of Long-term Debt (due within 1 year)

An understanding of the various financial ratios and percentages available to the contractor from the balance sheet and statement of profit and loss will enable him to determine how he is using his financial resources and where the weaknesses and strengths are. This, in turn, will allow the contractor to take steps to properly utilize the financial resources available to him. In addition, the ratios and percentages will supply a reliable basis for comparison with competitors' business practices.

Acquiring working capital

Working capital can be increased by maintaining at the lowest possible level the investment in building, tools, trucks, furniture, and inventory. However, to preserve the ability to compete and to complete work as required, it is necessary to have adequate facilities to sustain employee morale and efficiency and the tools and rolling stock required to properly perform all job functions.

Inventory should be kept at the lowest possible level which will permit prompt supply of material to the job site. To accomplish this it is necessary to develop competent suppliers who will work with you on a consistent basis.

Though inventory is considered to be a current asset, it does require a cash investment which reduces the amount of cash available for operating the business in the same way that investment in fixed assets does.

Working capital may be increased by obtaining a long-term loan (in the neighborhood of 3 to 5 years). A loan for this period is considered a long-term liability; therefore, the net amount of cash obtained through the loan minus the amount due in 1 year or less will apply directly on the balance sheet to working capital (recall that we have defined working capital as the difference between current assets and current liabilities).

Another method of calculating working capital is to take net worth minus fixed assets plus long-term liabilities:

$$\text{Working capital} = (\text{net worth} - \text{fixed assets}) + \text{long-term liabilities}$$

If working capital is the pivot point about which the wheel of business revolves, then net worth is the axle which supports the entire financial load of the business. Without sufficient net worth the business is doomed to failure because without sufficient net worth banks will not loan funds for working capital and the purchase of tools and materials. Net worth represents the value of the business, and the amount of net worth is determined by subtracting the total liabilities from the total assets.

Total assets are divided into the following three categories: (1) current assets, (2) other assets, and (3) fixed assets. Other assets, which are not as liquid as current assets, are:

1. Land held for investment
2. Cash surrender value of life insurance
3. Investment in stocks or bonds

Fixed assets consist of property and equipment such as:

1. Raw land
2. Buildings and improvements thereon
3. Construction equipment

4. Vehicles
5. Office equipment
6. Furniture and fixtures

Total liabilities consist of the following: (1) Current liabilities, (2) Long-term debt (due in over 1 year), and (3) Commitments. (*Commitments* are obligations such as cosigned contract bonds, contracts, or any other financial responsibilities.)

Leverage of working capital

Because the ratio of achievable sales volume to working capital ranges from 6:1 to 10:1, a contractor should find it attractive to use borrowed capital obtained on a long-term basis to increase working capital. In fact, it can be shown that the contractor can increase profits considerably by using this method. The contractor planning to borrow funds for working capital should be familiar with three basic methods used by banks to calculate interest. These methods are:

1. *Add-on interest:* Using this method, the bank would simply take the amount of the loan and add the total interest at the agreed-upon percentage for the entire period of the loan, and then divide by the applicable number of months to determine the monthly payment. For example, for a $25,000 loan for 5 years at 12 percent:

Loan amount	$ 25,000
Plus interest: 5 × 12% × $25,000	15,000
Total loan amount/number of months	$40,000/60
Monthly payment	$ 666.66

2. *Discount interest:* Using this method the bank would calculate the interest as for add-on interest and deduct the amount of interest from the proceeds of the loan. This is the most expensive type of loan because to obtain the amount of $25,000 for 5 years at 12 percent the contractor would have to borrow approximately $62,500:

Loan amount	$ 62,500
Minus interest: 5 × 12% × $62,500	37,500
Total net proceeds	$100,000
Monthly payment	$ 1,666.67

3. *Simple interest:* Using this method, the borrower pays interest on only the amount of the loan outstanding at any given time, with the amount of interest paid each month decreasing as each payment is

made. This loan will have a monthly amortization schedule, but in summary a 5-year, $25,000 loan at 12 percent will appear as:

Loan amount	$25,000.00
Monthly payment	$ 556.12
Total amount of payments: 556.12 × 60	$33,672.20
Interest paid	$ 8,672.20

The simple interest loan would save the contractor $28,827.80 over the discount method and $6,327.80 over the add-on interest method.

It should be possible for the contractor with sufficient collateral to borrow at the simple interest rate, but it is vital that the contractor thoroughly understand the differences between the types of loans commonly granted by banks. It will be in the contractor's best interest to shop several banks when seeking to borrow money.

Table 3.1 gives an example of the results of the leverage possible with an infusion of borrowed working capital in a business with an annual sales volume of $200,000 and working capital of $25,000, or sales of 8 times working capital. The table shows the statement of profit and loss before and after application of the borrowed funds. Table 3.2 shows information from the balance sheet before and after application of the funds.

TABLE 3.1 Statement of Profit and Loss Before and After Leveraging Working Capital

Before Applying $25,000 Loan to Working Capital:		
Material cost	$ 70,000	35% of sales
Labor cost	54,000	27% of sales
Direct job expense	18,000	9% of sales
Prime cost	$142,000	71% of sales
Overhead cost	50,000	25% of sales
Total cost	$192,000	
Profit	$ 8,000	4% of sales
Total sales	$200,000	
After Applying $25,000 Loan to Working Capital:		
Material cost	$140,000	35% of sales
Labor cost	108,000	27% of sales
Direct job expense	36,000	9% of sales
Prime cost	$284,000	71% of sales
Overhead cost	92,000	23% of sales
Profit	$ 24,000	6% of sales
Total sales	$400,000	

TABLE 3.2 **Balance Sheet before and after Leveraging Working Capital**

Information from Original Balance Sheet			
Current assets	$42,800	Current liabilities	$17,800
Fixed assets	10,700	Net worth	35,700
Total	$53,500	Total	$53,500
Working capital: $42,800 − 17,800 = $25,000			
Information from Balance Sheet after $25,000 Loan			
Current assets	$67,800	Current liabilities	$17,800
Fixed assets	10,700	Long-term liabilities	25,000
		Net worth	$35,700
Total	$78,500	Total	$78,500
Working capital: ($35,700 − $10,700) + $25,000 = $50,000			

In the example the contractor would be paying about 1 year's profit for a $25,000 loan over a 5-year period, but he could triple his profit and increase his overhead by 85 percent, while reducing his overhead adder by 2 percent to become more competitive. His payoff on the loan at simple interest would be $556.12 per month for the 5-year period.

To accomplish the increased volume of work the labor force would be doubled, but at the same time the owner could increase his salary and still leave double the profit in the business upon which to grow and increase the working capital. It would be necessary to operate the larger volume of sales with the same amount of inventory and fixed assets.

The cost of the loan would probably be somewhat higher than the bank's prime rate but the contractor doing a small volume of business presents a greater risk for the lender. It would appear that such a loan would be advantageous for all concerned provided the contractor maintains control over all the factors of the business. The important controls which must be exercised are:

1. Keeping the sales volume to no more than 10 times working capital
2. Collecting accounts receivable on a current basis
3. Ensuring payment of the loan by making a profit
4. Controlling overhead costs to recognized proper ratios
5. Maintaining correct amounts of investment in fixed assets and inventory

In addition to the long-term loan the contractor should obtain an open line of credit with the bank to enable him to finance temporary peaks in sales volume. This would be on a short-term basis to provide working capital as needed for special requirements.

Almost every contract requires a retention of at least 10 percent of total billing. Since this retention comes out of accounts receivable and is, therefore, part of working capital, it is incumbent on the contractor to replace this amount of cash retained by the customer.

One method of doing this is to bill the job in advance as much as possible. Another method would be to assign the accounts receivable to the banks as collateral for a loan against accounts receivable.

The importance of achieving profitable sales in accordance with allowances permitted by the amount of working capital cannot be overemphasized. Therefore, working capital is the limiting factor of the business. As mentioned above a long-term loan, short-term loan, and receivables financing by a bank are sources of increases to working capital. Other sources would be outside sales of stock in the business or perhaps private investment as a partner.

Table 3.3 is a typical example of the manner in which you may determine changes in your company's financial position at the end of any given accounting period. This information is all obtained from a comparison of your balance sheet at the beginning of the period to your balance sheet at the end of the period.

Control of working capital

To ensure the ongoing growth and success of the business it is imperative that management control the financial affairs of the business by doing the following:

1. Ascertain the sales volume available to the company in the market available to you and in the type of work you are familiar with. Set up a proper amount of working capital to support that volume of sales. Set an obtainable goal of a volume of sales based on the company's margins of overhead and profit. In other words, control sales and working capital.
2. Control the elements of working capital as follows:
   a. *Net worth:* Have sufficient capital invested in the business.
   b. *Profit:* Make a profit in the operation of the business.
   c. *Fixed assets:* Allocate the investment of capital in proper proportions as follows: tool inventory allowance, 5 percent of net worth; material inventory, 10 percent of net worth; furniture and fixtures, 10 percent of net worth; building rent, 11 percent of net worth; truck rental, 14 percent of net worth.
   d. *Long-term liabilities:* Consider borrowing capital for at least a 5-year payout to shore up working capital and expand sales.

TABLE 3.3 Calculation of Period-to-Period Change in Working Capital

Source of working capital		
From operations:		
Net earnings		$ 55,650
Depreciation (no cash outlay)		9,800
Working capital from operations		$ 65,450
Proceeds from long-term borrowing		$ 58,450
Cash value of life insurance		31,100
Sale of land held for investment		7,600
Sale of fixed assets		3,800
		$100,950
Gross increase in working capital		$166,400
Decreases in working capital:		
Additions to equipment	$39,250	
Payment on long-term debt	12,800	
Increased cash value of life insurance	3,050	
		$ 55,100
Current increase in working capital		$111,300
Working capital at beginning of period		99,300
Total working capital at end of period		$210,600
Increases in current assets:		
Cash and certificates of deposit		$ 53,100
Accounts receivable		100,300
Underbillings		29,000
Other current assets		13,100
		$195,500
Increases in current liabilities:		
Accounts payable	$30,000	
Overbillings	13,450	
Income taxes	38,300	
Other current liabilities	2,450	
		$ 84,200
Working capital (current assets − current liabilities)		$111,300

Operating the business within the limits imposed by the financial condition of the company is absolutely necessary to assure the successful continuation and growth of the business. This requires perpetual examination of accounting reports and the willingness to make corrections in the operations promptly when required to maintain the proper balance of the financial elements.

Financial Structure of the Business

There are three periodic financial reports that are required for financial management of a business:

1. Summary of income and expense
2. Summary of overhead expense
3. Balance sheet

These reports must be completed and available to management on a timely basis and every manager must be familiar with and understand each of these reports. With the computer programs and equipment available today it is possible for every business, regardless of size, to have timely financial reports which enable management to properly control this aspect of the operation.

The summary of income and expense, sometimes referred to as the statement of profit and loss, must be issued every month for proper control. A timely report allows management to make comparisons to the budget. In most cases the bottom line, or profit figure, is of primary importance but the sales figure is the controlling figure. All other financial considerations are dependent on the sales figure because without achieving budgeted sales it is impossible to control the other expense items and make a profit. Table 3.4 illustrates a typical summary of income and expense.

TABLE 3.4

XYZ Electric Company
Summary of Income and Expense
for Period Month of __________

		% of Sales
Sales	$1,904,831.60	100.00
Less cost of sales:		
Material	$ 696,778.29	36.5
Labor	521,397.91	27.37
Payroll tax	157,028.40	8.24
Direct job expense	61,473.14	3.23
Subcontract expense	59,799.58	3.14
Prime cost (total sales cost)	$1,496,477.32	78.56
Gross profit	$ 408,354.49	21.44
Less overhead costs	369,111.49	19.38
Operating profit	$ 39,242.79	2.06
Less income taxes	14,967.73	.79
Net profit	$ 24,275.06	
Retained earnings beginning of period	$ 215,000.00	
Retained earnings to date	$ 239,275.06	

Statistics

The National Electrical Contractors Association (NECA) has done an excellent job of publishing annual statistical information every 2 years for the electrical contracting industry. An income and expense report for their membership by payroll group for the year 1990 is given in Table 3.5.

It is imperative that every business manager compare their individual business statistics with industrywide statistics in order to evaluate how they compare from the standpoint of competition.

The percentages shown in Table 3.4 are based on sales. The percentage adders used in pricing estimates, material, or labor are based on prime cost, so it is necessary to recalculate these adders as illustrated in Table 3.6. Note that to recover overhead and earn a profit it is necessary to add the percentages of total cost for overhead and profit as indicated in Table 3.6 to the estimated prime cost.

Tables 3.5 and 3.6 can be used to illustrate a typical bid. It is assumed that a contractor in group 4 is bidding on a $1 million project; his average bid would break down as follows:

Material cost	$ 374,700	actual cost
Labor cost	272,700	actual cost
Labor adder: 34.69% of labor	94,600	
Direct job expense	36,200	actual cost
Subcontract	34,200	actual cost
Total prime cost	$ 812,400	
Overhead: 18.8% of prime cost	152,731	
Profit: 4.29% of prime cost	34,869	
Contract estimate	$1,000,000	

The estimated material cost would be arrived at from vendor quotes on the Bill of Material. The labor estimate would be developed from the contractor's established labor units and applicable rates. The direct job expense cost would be determined by the contractor's past job experiences. The subcontract cost is the result of subcontractors' bids on the job. Labor adder, which includes payroll taxes and benefits, and overhead and profit are actual add-ons as a percentage of prime cost.

Overhead

It is interesting to compare the group 3 and group 4 contractors' overhead and profit percentages noted in Table 3.5. The group 3 contractor would require 3.62 percent more adder onto prime cost to cover his overhead and profit. On the typical bid as illustrated above, the group 3 contractor would apparently be $29,409.00 higher than the group 4

TABLE 3.5 Summary of Income and Expense for NECA Membership, 1990

	Group 1: annual direct payroll under $250,000		Group 2: annual direct payroll $250,000 to $500,000		Group 3: annual direct payroll $500,000 to $1,000,000		Group 4: annual direct payroll $1,000,000 to $2,500,000	
	Average dollars	Percent of sales	Average dollars	Percent of sales	Average dollars	Percent of sales	Average dollars	Percent of sales
Total sales	711,856	100.0	1,476,811	100.0	2,830,161	100.0	6,042,106	100.0
Prime cost:								
Material	258,676	36.34	569,621	38.57	1,064,257	37.60	2,264,090	37.47
Direct labor wages	149,533	21.01	361,549	24.48	715,642	25.29	1,647,695	27.27
Labor adder	49,141	6.90	108,583	7.35	259,230	9.16	571,538	9.46
Other direct job expense	16,895	2.37	47,069	3.19	77,914	2.75	218,742	3.62
Subcontract expense	29,532	4.15	47,911	3.24	116,489	4.12	206,399	3.42
Total prime cost	503,777	70.77	1,134,734	76.84	2,233,537	78.92	4,908,464	81.24
Gross income*	208,079	29.23	342,077	23.16	596,624	21.08	1,133,642	18.76
Total overhead expense	175,471	24.65	286,506	19.40	490,369	17.33	922,989	15.26
Operating profit	32,608	4.58	55,571	3.76	106,255	3.75	210,653	3.49
Interest expense	5,034	.71	9,883	.67	13,759	.49	30,665	.51
Nonoperating expense	5,514	.77	7,509	.51	18,598	.66	39,690	.66
Net profit before tax	33,078	4.65	53,194	3.60	111,092	3.93	219,678	3.64
Income taxes	9,278	1.30	13,180	.89	33,880	1.20	60,697	1.00
Net profit after tax	23,800	3.34	40,014	2.71	77,212	2.73	159,981	2.63

*Gross income = total sales − total prime cost.

TABLE 3.6 Recalculation of Overhead to Percent of Prime Cost

	Group 1		Group 2		Group 3		Group 4	
	Amount	Percent of prime cost	Amount	Percent of prime cost	Amount	Percent of prime cost	Amount	Percent of prime cost
Prime cost	$503,777		$1,134,734		$2,233,537		$6,042,106	
Overhead	$175,471	34.8	$ 285,506	25.3	$ 490,369	21.9	$ 922,989	18.8
Profit	32,608	6.5	$ 55,571	4.90	$ 106,255	4.8	$ 210,653	4.3
Sales	$711,856		$1,476,811		$2,830,161		$6,042,106	

contractor assuming that the material, labor, subcontract, and direct job expense were the same. Doing the same comparison between the group 1 and group 2 contractors would reveal a difference of 11.15 percent of prime cost.

Common practice in the electrical industry dictates that an overhead adder of 15 percent and a profit adder of 10 percent of prime cost may be used to price Change-Order work. It is obvious from Tables 3.5 and 3.6 that all four of the groups listed would lose money using these adders.

These illustrations highlight how necessary it is for each manager to know the financial statistics of his business and apply them properly to each bid. The illustrations are broad averages of many contractors across the nation and should not be used other than as guidelines. Each contractor must determine the adders that should apply to his particular business based on his financial condition and mode of operation.

Table 3.7 gives data published by NECA for the year 1990, showing overhead costs in the four groups listed in Tables 3.5 and 3.6.

The data shown in Table 3.7 conform in summary to that shown in Table 3.5, and as indicated by the table all percentages are a percent of sales. The data on overhead are useful to provide a basis for the preparation of a budget for projection of allowable overhead expenses and a forecast of the sales required for the following year.

Every contractor must know his own costs and operate his business accordingly, but it is imperative that he not only know his own costs and how they compare with the competition but know also how each item compares as a percentage with other data from his own company.

Overhead costs vary widely among companies doing different types of construction work. For example, a contractor doing only small inside electrical work will have a larger overhead percentage than the contractor doing large industrial-type jobs. Therefore, it is important to know the type of business to which you are comparing in order to get an accurate comparison of data.

TABLE 3.7 Summary of Overhead Expenses

	Group 1: annual direct payroll under $250,000		Group 2: annual direct payroll $250,000 to $500,000		Group 3: annual direct payroll $500,000 to $1,000,000		Group 4: annual direct payroll $1,000,000 to $2,500,000	
	Average dollars	Percent of sales	Average dollars	Percent of sales	Average dollars	Percent of sales	Average dollars	Percent of sales
Total overhead	175,471	24.65	286,538	19.40	490,362	17.33	922,989	15.28
Advertising and promotion	2,887	0.41	3,433	0.23	7,542	0.27	13,705	0.23
Autos and trucks	10,066	1.41	15,911	1.08	21,051	0.74	37,141	0.61
Bad debts	2,524	0.35	4,124	0.28	4,923	0.17	16,075	0.27
Charitable contributions	599	0.08	976	0.07	1,697	0.06	2,969	0.05
Collection expense	75	0.01	333	0.02	16	0.00	175	0.00
Depreciation	9,149	1.29	14,800	1.00	23,158	0.82	41,725	0.69
Dues and subscriptions	1,907	0.27	2,513	0.17	5,776	0.20	5,333	0.09
Education expense	457	0.06	645	0.04	1,255	0.04	1,820	0.03
Overhead, employee benefits	6,521	0.92	8,220	0.56	9,150	0.32	22,164	0.37
Freight and express	173	0.02	199	0.01	523	0.02	925	0.02
Heat, light, water	2,770	0.39	3,006	0.20	5,606	0.20	8,545	0.14
Insurance, general	8,262	1.16	14,516	0.98	22,190	0.78	32,259	0.53
Overhead, employee workmens compensation	4,208	0.59	5,615	0.38	6,263	0.22	18,066	0.30
Employer's liability insurance	1,779	0.25	3,741	0.25	4,523	0.16	7,149	0.12
Legal and accounting	4,048	0.57	7,219	0.49	11,695	0.41	27,329	0.45
Miscellaneous	1,879	0.26	2,790	0.19	3,194	0.11	7,318	0.12
Office supplies	5,076	0.71	7,205	0.49	11,229	0.40	21,140	0.35
Pension and profit sharing	5,190	0.73	6,149	0.42	17,543	0.62	32,072	0.53
Rent	9,578	1.35	12,993	0.88	17,283	0.61	39,041	0.65
Repairs and maintenance	2,214	0.31	4,613	0.31	6,606	0.23	12,032	0.20
Plan, bonds, engineering expense	817	0.11	874	0.06	1,829	0.06	5,543	0.09
Overhead, salaries and wages	69,407	9.75	132,215	8.95	242,355	8.56	457,240	7.57
Shop supplies and expenses	1,246	0.18	1,534	0.10	3,517	0.12	6,370	0.11
Small tools	1,259	0.18	1,814	0.12	3,583	0.13	9,060	0.15
Taxes and license fees	3,730	0.52	4,839	0.33	8,754	0.31	17,001	0.28
Taxes, overhead, payroll	10,724	1.51	11,603	0.79	21,598	0.76	34,491	0.62
Telephone and postage	3,932	0.55	6,510	0.44	9,630	0.34	18,905	0.31
Travel (administrative)	1,641	0.23	3,407	0.23	4,342	0.15	10,639	0.18
Other expenses	3,355	0.47	4,656	0.32	13,716	0.48	13,579	0.22

SOURCE: NECA.

The accountant refers to overhead cost as "general and administrative expense," but in the electrical construction industry these costs are generally known as "overhead." As a rule it is not practical to charge these expenses to a particular part of the work and as a result they must be absorbed as an adder to prime cost.

The overhead costs are fixed costs and they accrue whether or not any sales are made. The costs are controllable within certain limits but once management has committed to a purchase or lease of real estate, for example, these costs are seldom subject to change.

Since the overhead costs must be met before any profit can be made, or for that matter before the administrative payroll can be met, the financial management of the proration of overhead costs to each job accomplished is of primary importance to the success of the business. While the proration of overhead to prime cost is not difficult, it does require a methodical approach to ensure its financial success. Each job must bear its proportionate share of the total overhead expense and a sufficient volume of sales must be generated to absorb this expense.

Application of overhead. As an example of application of overhead, consider the figures for group 3 in Table 3.6. Here it is necessary to cover overhead costs of $490,369 by a volume of sales of $2,830,161; the overhead is 21.95 percent of prime cost.

The simplest way to cover this overhead is by adding 21.95 percent to the prime cost of every job, provided that the prime cost of every job includes 47.65 percent material, 32.04 percent direct labor, and 3.49 percent direct job expense. It is unlikely that this exact relationship of prime costs will occur very often, but the one thing that will occur on every job is man-hours of labor.

Budgeting of overhead. On many jobs a great deal of material is furnished by the owner, so the amount furnished by the contractor is reduced. Therefore, in order to ensure coverage of overhead cost and proper proration of overhead, it appears necessary to add overhead costs on the basis of man-hours of labor. Overhead cost per man-hour is called *labor burden.*

$$\text{Labor burden} = \frac{\text{overhead cost}}{\text{man-hours labor}}$$

This procedure is especially valid for service work, where it is essential to develop a resale rate per man-hour for labor. It is assumed that a yearly budget for a service department which operates 10

TABLE 3.8 Service Budget and Determination of Overhead Rate per Man-Hour

	Amount	Percent of sales
Material	$245,000	28
Labor: 20,000 man-hours @ $14.00	280,000	32
Payroll tax (31% of labor)	87,500	10
Truck expense (9% of labor)	26,250	3
Prime cost	$638,750	73
Overhead	183,750	21
Profit	52,500	6
Total sales	$875,000	100

Under these conditions the labor burden would be:

$$\frac{\$183{,}750 \text{ (overhead)}}{20{,}000 \text{ man-hours}} = \$9.19 \text{ per man-hour}$$

trucks with a journeyman pay scale of $14.00 per man-hour would be as shown in Table 3.8.

The service rate per man-hour based on applying all overhead on labor would be:

Labor	$14.00
Payroll tax (31%)	4.34
Truck expenses (9% of labor)	1.25
Prime rate	$19.59
Overhead	9.19
Profit (10% of prime cost)	1.96
Labor rate	$30.74

In order to be as competitive as possible it would probably be advisable to reduce the rate per man-hour by applying some of the overhead cost to material. By adding 25 percent to material cost a new amount will be developed: $245,000 × 0.25 = $61,250. This amount may be subtracted from the overhead amount of $183,750 to leave $122,500; this can be applied to the 20,000 man-hours of labor, which results in a labor burden of $6.12 per man-hour instead of the previous $9.19, or a reduction of $3.07 per man-hour and a service labor rate of $27.67 per hour.

The amount of the service rate is critical because the customer is familiar with the labor rate from a competitive standpoint. It pays to recognize the importance of the rate per man-hour as long as the necessary dollars of overhead are covered by a margin on material.

Balance sheet

The balance sheet is a statement of financial condition—of assets, liabilities, and net worth as of a certain date. The published form of a balance sheet is usually prepared at the end of the company's fiscal year in conjunction with the summary of income and expense and the summary of overhead expenses; however, the balance sheet is required in order to evaluate working capital and therefore should be prepared, preferably monthly, but in any case often enough to allow the manager to exercise proper control.

Table 3.9 illustrates the asset side of a typical balance sheet for a medium-size contractor and Table 3.10 illustrates the liability side for the same contractor.

TABLE 3.9

XYZ Company
Balance Sheet
For Period Fiscal Year End

Assets

Current assets:		
Cash		$ 47,100
Certificates of deposit		100,000
Note Receivable		7,300
Accounts Receivable including retention of $135,100		265,350
Unbilled amounts on completed contracts		10,800
Material and supplies inventory		19,650
Costs and estimated earnings on contracts in progress in excess of related billings		36,750
Prepaid expenses and other current assets		9,450
Total current assets		$496,400
Other assets:		
Land held for investment		5,000
Cash surrender value of life insurance net of policy loans of $31,100		1,800
Total other assets		$ 6,800
Fixed assets:		
Land		$ 5,300
Building and improvements		57,600
Construction equipment		82,500
Vehicles		20,500
Office equipment		7,250
Total fixed assets	$173,150	
Less: Accumulated depreciation	65,600	
		$107,550
Total assets		$610,750

TABLE 3.10

XYZ Company Balance Sheet For Period Fiscal Year End	
Liabilities and Stockholders' Equity	
Current liabilities:	
Accounts payable, including retention of $72,000	$137,750
Accrued payroll, payroll taxes, union benefits, and amounts withheld from employees	4,000
Additional costs on completed contracts	4,300
Amounts billed in excess of costs and estimated earnings on contracts in progress	66,450
Other current liabilities	5,500
Income taxes:	
Current	7,500
Deferred	52,800
Current portion of long-term debt	7,500
Total current liabilities	$285,800
Long-term debt	$ 91,350
Stockholders' equity—net worth:	
Common stock, par value $10 per share; authorized, 10,000 shares; issued and outstanding, 1,850 shares	$ 9,250
Additional paid-in capital	5,600
Retained earnings	218,750
	$233,600
Total liabilities and net worth	$610,750

This example indicates that *normal* working capital equals current assets minus current liabilities. In this case,

$$\text{Working capital} = \$496{,}400 - \$285{,}800$$
$$= \$210{,}000$$

It must be recognized that this statement shows long-term debt. Therefore, long-term debt must be recognized as part of working capital, as follows:

$$\text{Working capital} = (\text{net worth} - \text{fixed assets}) + \text{long-term debt}$$
$$= \$233{,}600 - \$102{,}550) + \$91{,}350$$
$$= \$131{,}050 + \$91{,}350$$
$$= \$222{,}400$$

While this example shows a larger working capital (larger by $11,800), it must be recognized that the long-term debt must be paid off, and when this occurs there must be continuing profits added to the net worth, and continuing depreciation of fixed assets in order to sustain needed working capital. Management's control of the financial aspects of the business is almost entirely a function of understanding the meaning of the various financial ratios and acting to keep all the financial ratios within allowable limits. A *ratio* is the relationship in quantity, amount, or size between two or more things, and financial control is a process of maintaining a proper balance between all the financial functions of the business.

We have indicated in previous sections the functions that must be controlled and how that is accomplished as follows:

1. Working capital = current assets − current liabilities

and also

Working capital = (net worth − fixed assets) + long-term debt

2. Fixed assets = tools and equipment (5% net worth)
   trucks and autos (14% net worth)
   furniture and fixtures (10% net worth)
   land and buildings (11% net worth)
3. Material inventory = 10% net worth
4. Sales = 10% × net worth
5. Net worth = total assets − total liabilities
6. Overhead = (sales − prime cost) − profit

Bases for the financial ratios have already been established for the purpose of management control, but it is advisable to further examine the total spectrum of financial ratios. All the financial ratios are taken from data furnished by the balance sheet.

To use financial ratios intelligently you must:

1. Develop an understanding of the terms used on a balance sheet and the mechanics of its makeup.
2. Establish the relative *values* of the economic components of a company.
3. Develop an understanding of the important relationships within the financial structure of the company.

The term itself, *balance sheet,* implies that we are equating the assets to the sum of the liabilities plus net worth. The two quantities must be equal.

Assets include	Current Assets Fixed Assets Company Investments

Liabilities include	Current Liabilities Long-term liabilities
Net Worth Includes	Common Stock Paid-in Capital Retained Earnings

Since the assets side and the liabilities side are equal, and since we are searching for a method of equating the components of the two sides, it is logical to envision the entire financial structure as a circle with a line down the center to equate two equal sides, with their components being smaller pieces of the pie, thus:

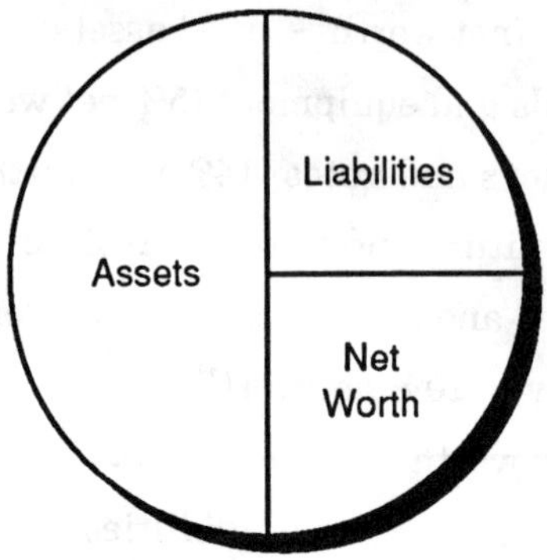

Current assets vs. fixed assets. Because it is generally necessary to keep the assets of the company available for meeting current obligations, it is generally desirable to keep the amount of assets invested in fixed (nonliquid) form to a minimum. Let us set 20 percent of total assets as a reasonable value.

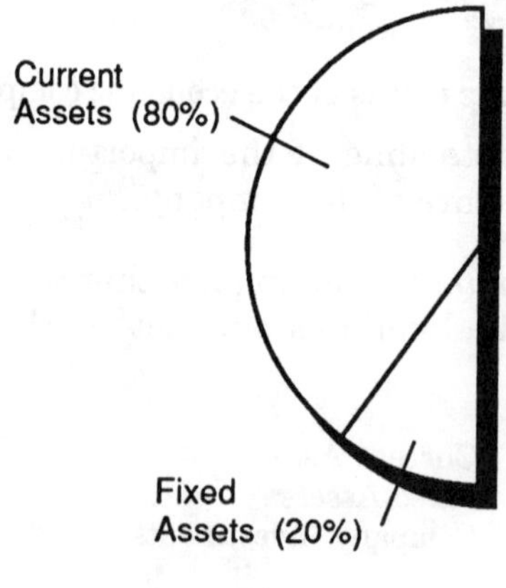

Liabilities vs. net worth. With the goal of supplying reasonable protection to creditors and at the same time ensuring that a company can weather a bad year, it is good practice to have the net worth of a company equal to its total liabilities.

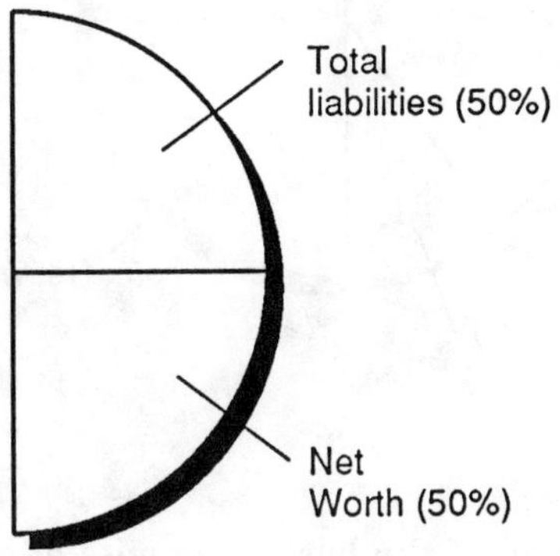

Current liabilities vs. fixed (long-term) liabilities. Current liabilities will generally form the bulk of the total debt. Let us assume that fixed liabilities (those due in more than 1 year) are 10 percent of the total liabilities:

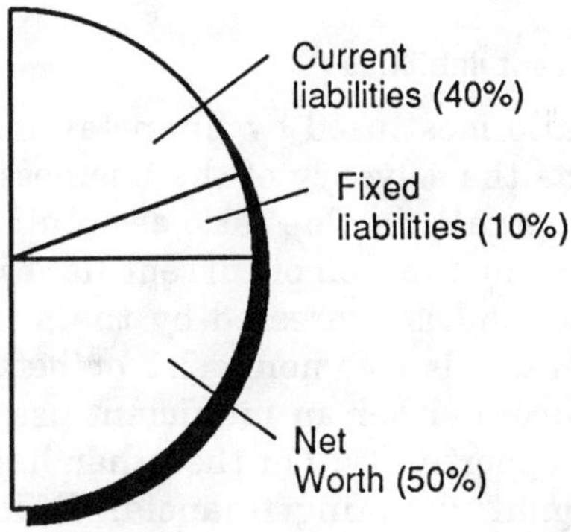

The overall picture. Putting all the individual components together, we can complete the circle and determine the overall relationships:

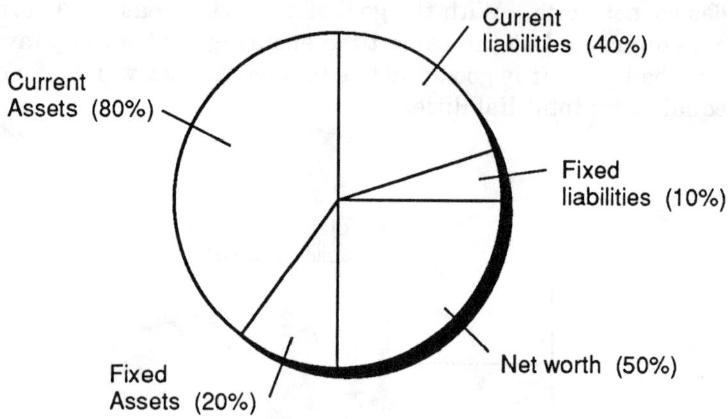

Financial Ratios

With an understanding of the basic elements of a balance sheet and the proportions of an ideal company, we are ready to investigate the meaning of some of the frequently used financial ratios and then establish a normal value for later comparative use. It will be useful to note before we begin the discussion that any ratio can be expressed as a percentage by converting the ratio to a decimal fraction and then multiplying by 100. Thus, a ratio of 1:10 expressed as a decimal fraction is 0.10, and $0.10 \times 100 = 10\%$.

Current assets to current liabilities

This may be the ratio most used by financial analysts, the reason being that it reflects the solvency of the business and its ability to finance day-to-day operations. The ratio as obtained by dividing the sum of current assets by the sum of current liabilities, is referred to as the *current ratio,* and is expressed by the formula $CA/CL = R$. The ideal for this ratio is commonly 2:1 or better. An extremely high ratio may indicate either an inefficient use of funds or a possible lost business opportunity. On the other hand, a ratio of less than 1.5:1 may signal upcoming financial difficulties from overextension of the business.

The company's working capital is represented by the sum of the current assets less the sum of the current liabilities. Working capital determines the volume of sales which can be supported by the company and should be no less than 10 percent of the total sales volume.

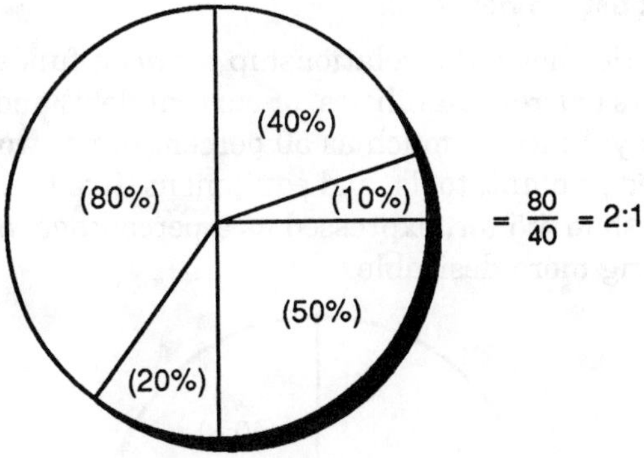

$$\text{Current ratio} = \frac{\text{current assets}}{\text{current liabilities}}$$

Fixed assets to net worth

This ratio is obtained by dividing the sum of fixed assets before depreciation, as shown on the balance sheet, by net worth. The accepted norm in the electrical industry for the amount of investment in fixed assets is from 30 percent to not over 50 percent of net worth. Investment of less than 30 percent of net worth in fixed assets could result in an inability to supply needed tools or rolling stock to the job, which will cause excessive labor costs to be accrued. Another result of inadequate investment in fixed assets could be poor facilities, which can cause inefficiencies in operation of the office or warehouse. Overinvestment suggests lost revenue because of fixed assets sitting idle in the storage yard which reflects in reduced working capital.

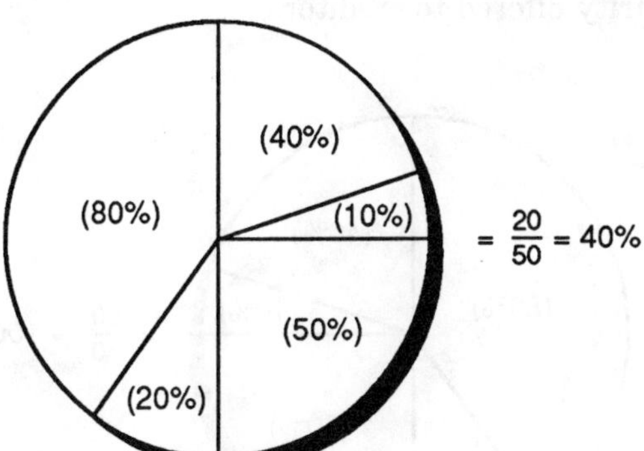

$$\text{Ratio} = \frac{\text{fixed assets}}{\text{net worth}}$$

Current debt to net worth

This ratio shows the relationship between funds owed to "temporary" creditors (current liabilities, or current debt) and the total value of the company. Since as much as 50 percent of net worth may prudently be invested in plant, tools, and equipment (fixed assets), this ratio should be less than 4:5 (or, expressed as a percentage, 80%), with a lower figure being more desirable.

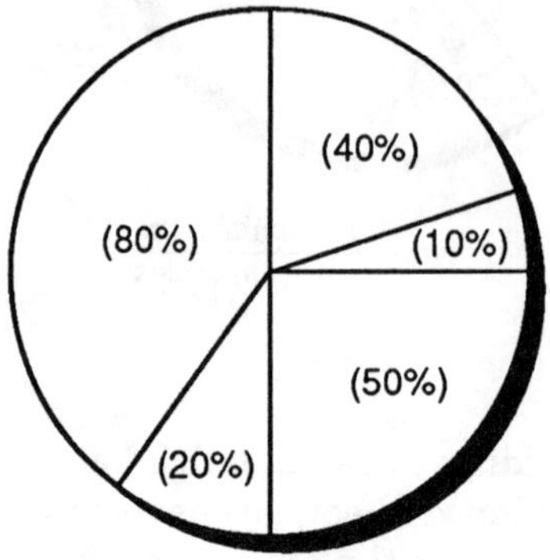

$$= \frac{40}{50} = 80\%$$

$$\text{Ratio} = \frac{\text{current debt}}{\text{net worth}}$$

Total debt to net worth

This ratio shows the relationship between the total of all obligations owed to creditors (both short and long term) and the total value of the company (net worth). If a company desires to be able to guarantee its obligations to all its creditors independent of any receivables it may have, this ratio must be 1:1 or less. Obviously, the smaller this ratio, the more security offered to creditors.

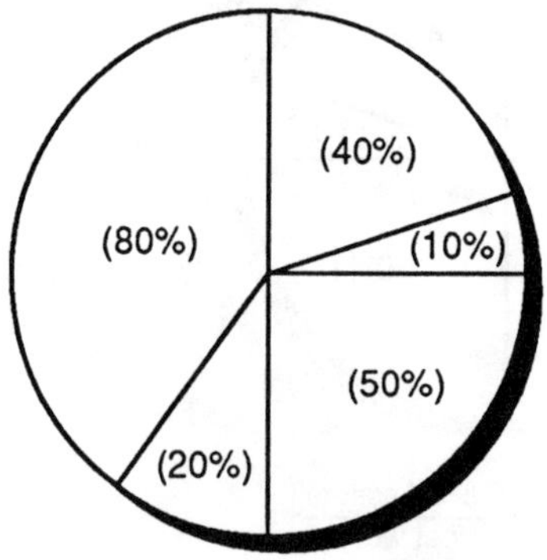

$$= \frac{50}{50} = 100\%$$

$$\text{Ratio} = \frac{\text{total debt}}{\text{net worth}}$$

Gross revenue to working capital

This ratio ties into the current ratio. If capital is turning at a low rate, sales are below normal, or a company has more current items than it needs and should be looking for more business or for other investment opportunities, then this ratio will be too low. The comfortable range in the electrical industry is a ratio of between 6:1 to 8:1. With efficient management and field operations along with prompt collection of accounts receivable, the company can operate successfully with a ratio of up to 10:1. A ratio higher than 7:1 is preferred. The formula is

$$\text{Ratio} = \frac{\text{gross revenue}}{\text{working capital}}$$

Developing a value for working capital

If we refer once again to the pie chart we have been working with (see under "The Overall Picture"), we can see that for "Ideal Electric," where current assets = 80 and current liabilities = 40, the value for working capital would be the difference between the two:

$$\begin{aligned}\text{Working capital} &= \text{current assets} - \text{current liabilities}\\ &= 80 - 40\\ &= 40\end{aligned}$$

Long-term liabilities to working capital

Because working capital can be increased directly by obtaining funds through long-term liabilities and using the funds to supplement current assets, it is most important to ensure that when the long-term liability must be paid off the corporation's activity will not be restricted because of a drop in working capital. As a guide, NECA recommends a value for this ratio below 1:4 (when expressed as a percentage, below 25 percent).

Net worth to current assets

This ratio is diagnostic of underutilization or overutilization of the net worth as current assets are increased due to business activity. An optimum range for this ratio when expressed as a percentage is from 50 to 75 percent. Values below 50 percent indicate that the company is working its resources very hard. On the other hand, values in excess of 75 percent indicate that the company may not be working its resources hard enough.

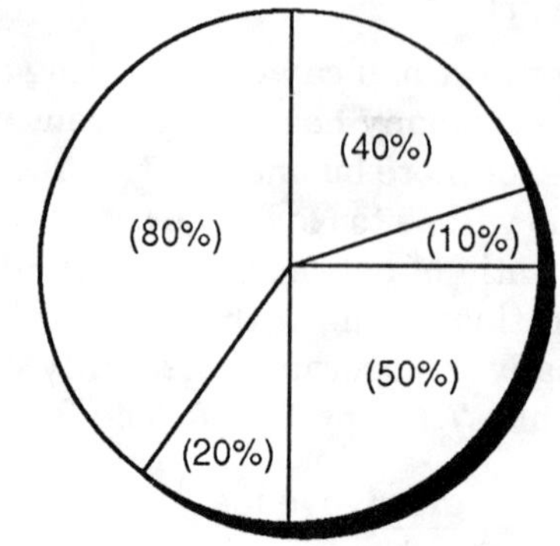

$$= \frac{50}{80} = 63\%$$

$$\text{Percentage} = \frac{\text{net worth}}{\text{current assets}} \times 100$$

Optimum range: 50 to 75%

Profit before tax to net worth

The final test of a company is the return the company earns on the value represented by its net worth. An analysis of the percentage return and its trend as compared with those of other companies and industries can indicate whether the company's net worth is prudently invested. Given the financial condition of most investments today a ratio of income (before tax) to net worth of 1:10 (10 percent) would be considered satisfactory. In view of the high-risk nature of the electrical contracting business, a value of 25 to 35 percent is a reasonable goal.

Profit before tax to working capital

This ratio is useful in measuring the performance of a contractor who either does not use all his net worth in the operation of the business or has obtained a substantial portion of his operating funds from long-term debt. Because of the need to repay long-term debt, a ratio of 3:10 to 4:10 (percentage of 30 to 40 percent) should be reasonable.

Profit before tax to net sales

The ratio of net income to net sales is another measure commonly used to compare the performance of one company to that of other companies or industries. However, this measure can be misleading when it is used to make comparisons to other industries. The relative stability, risk factor, and capital requirement per sales dollar must be considered when making any comparison. Obviously, the higher the percentage value for this ratio the better, with 8 to 12 percent considered desirable.

Usefulness of ratios and percentages

The principal value of financial ratios and percentages to a manager may be as a tool to compose financial conditions in his company to the industry standards. However, some of the ratios should also be used on a monthly basis as "action" controls.

Table 3.11 provides a summary of the accepted ranges for the critical ratios and percentages.

TABLE 3.11 Optimal Values for Critical Ratios and Percentages

Ratio	Ideal value
Current ratio	2:1
Fixed assets/net worth	40%
Current liabilities/net worth	80%
Total liabilities/net worth	100%
Gross revenue/working capital	10:1
Gross revenue/net worth	9.6:1
Debt/total capitalization	50%
Long-term liabilities/working capital	25%
Net worth/current assets	63%
Net income/net worth*	25%
Net income/working capital*	30%
Net income/gross revenue*	8%

*Net income before taxes.

A value for gross revenue

We have said that a well-run company may have a ratio of gross revenue to working capital of 10:1. If working capital is given a value of 40 (see under "Developing a Value for Working Capital"), then

$$\text{Gross revenue} = (10)\,(\text{working capital})$$
$$= (10)\,(40)$$
$$= 400$$

The ideal financial picture now appears as:

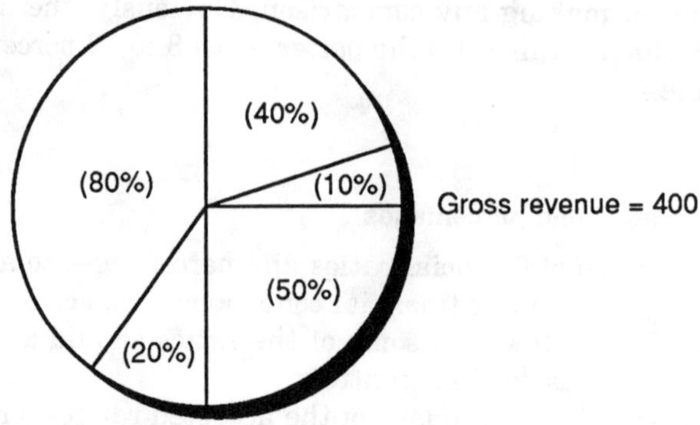

Gross revenue to net worth

This ratio indicates how well the net worth of a company is being utilized in developing sales. Assuming that profit increases with increasing sales, a ratio equal to or in excess of the "ideal" ratio is preferred—the "ideal" ratio being the ratio of the ideal gross revenue (400) to the ideal net worth (50).

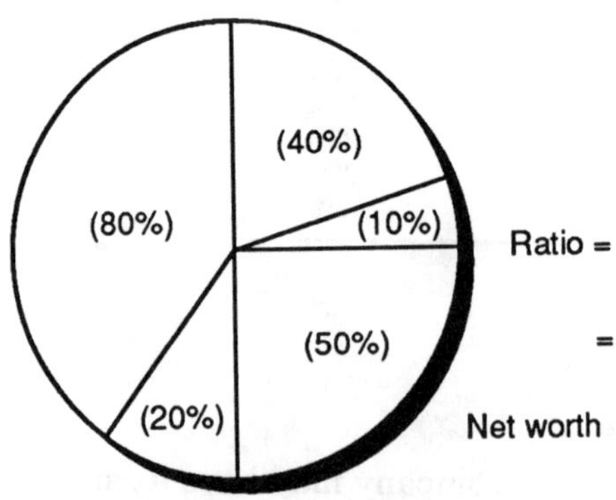

$$\text{Ratio} = \frac{\text{gross revenue}}{\text{net worth}} = \frac{400}{50} = 8:1$$

Some other important ratios and percentages that the manager may use to further control the financial stability of his company, along with their accepted norms, are described in the following sections.

Quick ratio

The *quick ratio* is so called because it refers to assets such as cash, marketable securities, and receivables which can be converted to cash quickly and compares these items with current liabilities. The stronger the company financially, the greater the ratio of quick assets is to current liabilities.

The strongest company will be able to cover all current liabilities with quick assets; in other words, the quick ratio will be 1:1 or greater (or, as a percentage, 100% or greater). The ratio is determined by taking the sum of the quick assets and dividing by the total of current liabilities. The formula is

$$\text{Ratio} = \frac{\text{quick assets}}{\text{current liabilities}}$$

Average collection period

Most electrical contracting business is conducted on a credit basis. The ideal collection period would be something less than 30 days, but in reality the industry experiences a collection period of from 40 to 50 days, which has been accepted as a standard. The average collection period can be determined in the following way: Determine the average monthly sales and collections by taking the gross sales for a year and dividing by 12 months, and then making the same calculation for gross collections. Then, by dividing average gross sales by average gross collections, a factor is developed which when multiplied by 30 days will give you the average collection period in days. The formula is [(gross sales/12)/(gross collections/12)] × 30 days = average collection period.

Since accounts receivables are an important part of the company's working capital and working capital is important to the sales volume, it is extremely important that the collection period be kept as low as possible.

Return on assets

Return on assets is expressed as a percentage representing profit before taxes as a percentage of total assets. This is a favorite measurement of return on investment but is subject to a lot of inconsistency due to the wide variation in the types of assets held by companies. For example, some companies have considerable sums invested in land which does not produce an immediate profit but is either necessary for future use of the company or is simply being held to take advantage of possible appreciation in land values. Return on assets is a useful measure of the year-to-year operation of the individual company but should not be construed as a good measure of comparison with other companies. The percentage is obtained by dividing the net profit before taxes by the total assets of the company and multiplying by 100.

Debt to total capitalization

This ratio measures debt, both current and long term, in relationship to total capital. Most financial analysts advise that the percentage derived from this ratio should not exceed a maximum of 50 percent. A company which shows a percentage greater than 66 percent may be overextending working capital, which could result in serious financial problems quickly. The percentage is calculated by dividing the total of current and long-term liabilities by total assets and then multiplying by 100.

Direct labor to gross revenue

This ratio will depend on the type of jobs the company has worked on during the current accounting period. For example, a hospital job will generally require a higher percentage of labor in relation to the sales value due to the complexity of the job. A warehouse job on the other hand will require a smaller amount of labor due to the simplicity of the job.

As the company performs more work, the percentage derived from this ratio becomes a valuable measure for management to determine the accuracy of estimates depending on the type of job which is being estimated. The averages reported for all types of work performed by a large group of electrical contractors in the United States have ranged from 31 to 33 percent. The formula for calculating this average is

$$\text{Percentage} = \frac{\text{direct labor}}{\text{gross revenue}} \times 100$$

Materials to gross revenue

This measure also varies somewhat depending on the type of job being constructed. The same large group of electrical contractors mentioned above report material used as a percentage of sales or gross revenue to be an average of 33 to 35 percent. This measure is of value to management not only to determine the validity of estimates but also to control material usage on the job site. The percentage is arrived at by dividing material cost by gross revenue and multiplying by 100.

General and administrative expense to gross revenue

Management has a prime responsibility to keep overhead and other administrative expenses to a level that allows the company to develop a reasonable profit. The constant measurement of general and administrative expense in relation to sales, and the control thereof, is imperative if the company is to survive. The ratio of general and administrative expense to the sum of gross revenue should, when converted to a percentage, range from 13 to 18 percent, with 18 percent being an extreme high.

The company's profit before tax is directly affected by the amount of overhead expense, and because the gross profit of electrical contractors is stringently controlled by the bidding process, the strict control of the overhead percentage is critical to the continued growth of the business.

Table 3.12 summarizes the accepted norms for a number of financial ratios and percentages. Along with Table 3.11, it provides a handy reference to the industry norms for these figures.

TABLE 3.12 Accepted Norms for Financial Ratios and Percentages

Ratio description	Ratio	Norm for ratio or percentage
Tools and construction equipment	Fixed assets − Tools / Net worth	5%
Trucks and autos (also rental)	Fixed assets − trucks / Net worth	14%
Furniture, fixtures, office equipment (also rental)	Fixed assets − fixtures / Net worth	10%
Land and buildings (also rentals)	Fixed assets − buildings / Net worth	11%
Material inventory	Inventory (current assets) / Net worth	10%
Sales/working capital	Sales / Working capital	10%
Sales/net worth	Sales / Net worth	9%
Profit/net worth*	Profit / Net worth	25%
Profit/working capital*	Profit / Working capital	30%
Profit/sales*	Profit / Sales	8%

*Before-tax profit.

Inventory

Numerous factors are involved when considering the relative merits of providing a material and tool inventory for the operation of an electrical contracting business. They are:

1. The contractor's prime responsibility in the successful performance of his installation work is to furnish the job with the proper material in the right quantity at the right time. This is accomplished by properly scheduling an accurate Bill of Material. The main problem with this is that it is very difficult for an estimator to take off material from plans with such a degree of accuracy that the Bill of Material doesn't require some adaptation to the conditions of the job—and this is generally not realized until the moment the material is required, so that the contractor must look to his warehouse supply for immediate delivery to the job.

2. The material costs involved in the estimate must conform to the prices at which it is purchased. This must be end-discount price or better in order to be competitive, and providing a reasonable warehouse stock of the staple items at the lowest possible price achieves the desired result.

3. The contractor's efficient performance requires the adaptation of standard installation methods using materials that conform to the warehouse stock. The field electricians must be trained to use these methods and materials.

4. Material inventory is considered a current asset, but as with the fixed assets, whatever values are involved reduce working capital by an equivalent amount.

Therefore the value of the inventory must be controlled to an amount equal to and *not exceeding* 10 percent of net worth. If a business has accurate inventory records it is possible to finance a large percentage of the inventory through a bank on a short-term loan which, if handled properly, may increase the functions of working capital as a long-term liability.

5. Nearly every job involves changes, additions, deletions, and modifications to the scope of the work. These affect the material requirements of the job, and for the most part demand immediate supply from the contractor's warehouse stock.

6. Contractors providing small-job service work require a shop material inventory in order to fill the immediate needs of the servicemen. This supply also provides the backup for small-contract jobs where the entire Bill of Material is dispatched to the job at one time. It is impractical for servicemen to rely on supply houses to fulfill their needs, because when they do too much time is lost making special trips to

pick up small items of standard material which must be available on short notice.

7. When an arrangement is worked out with a certain supply house to furnish all the material for a certain contractor's inventory requirements, the contractor is at a price disadvantage relative to the situation in which several suppliers are bidding to provide the material. Having a multiple source of supply ensures being in a better position to compete and make a profit.

8. The contractor must have his own tool inventory, which may be worked in conjunction with the material inventory where the personnel and space requirements may be compatible.

The value of the tool inventory may be equal to but not exceed 5 percent of net worth. The tool inventory values are part of the fixed assets on the balance sheet.

9. Expendable materials are an important part of the tool and materials inventory, but the individual items are not listed on the Bill of Material. The values represented by these items are equal to 0.20 percent of estimated labor cost, and the care and distribution of these items requires a custodial control of some kind in connection with the tool and material inventory.

Expendable materials are defined as items which are necessary to supplement the installation of materials, but do not remain as an integral part of the installation.

10. Packaging and storing facilities are essential to provide material and tool supply to the job. It is mandatory for the electricians on the job to be able to find the items they need without delay, and also to have proper facilities for transporting them to the point of use.

The contractor must not only stock and store these items, but they must be handled by the tools and material inventory personnel. There is no way a supply house could be expected to conform to this type of operation. Once the tools and materials have been delivered to the job site, security measures must be taken to protect them from theft.

11. Because of the many changes which occur during construction, there are many items of material left over when a job is completed, and all this material must be returned to the warehouse for processing. Some items may be returned to the supply house for credit; most, however, are highly usable but not in suitable condition for return to the vendor. It is essential for the contractor to have a warehouse facility to accommodate returns from the job.

12. In order to maintain accurate control over disbursements and pricing it is essential for the contractor to clear all materials through a central facility such as his own warehouse. It is very important to

maintain accurate cost records on every contract, and this involves careful paperwork controls which may be accomplished by channeling all charges and credits through one source.

The foregoing list shows conclusively that a contractor must have complete warehouse facilities to accommodate the following:

1. Material inventory
2. Tool inventory
3. Expendable material inventory
4. Packaging and storing facilities for both material and tools on job site

The financial values involved are the following:

Material inventory	10% of net worth
Tool inventory	5% of net worth
Expendable material	0.2% of estimated labor cost
Packaging and storing facilities (including truck and trailer storage)	2% of net worth

It is essential that accounting controls over the financial values represented by these items be maintained since working capital is diminished in equivalent value to them.

Budgeting

Abraham Lincoln said, "If we could first know where we are and whither we are tending, we could better judge what to do and how to do it." The budget for a business is the means of determining what to do and how to do it. The budget is the plan for coordinating all forces and directing them toward the attainment of a definite objective. It is the means of forecasting the following:

1. *Sales:* The volume anticipated for the ensuing year, when it will be achieved, by what departments, and by whom.
2. *Income:* The amount of income, the sources from which it will come, and how it will be produced.
3. *Expenditures:* The amount that will be spent for material, labor, supplies, expenses, and equipment and when it will be paid out.
4. *Financing:* The amount of financing necessary, when it must be provided, and when it can be liquidated.
5. *Profit:* The amount of profit and when it will be realized.

6. *Financial position:* Balance sheet figures, which, when projected into the future, will give effect to the operations of the business as estimated in the budget.

It is recognized in the electrical contracting business that there is no such thing as profit on any one job; that is, a wide variety of jobs of different sizes and types is necessary to meet the sales requirements of the business. Thus the most important responsibility of management is to achieve the sales volume requirements of the business and to include in the amount of each job a proper proportion of the total overhead costs and anticipated profit.

In order to achieve these goals it is necessary to plan ahead by preparing a budget for the ensuing year's operations. This budget should encompass all the financial elements of the business outlined above. The primary consideration in preparation of the budget is the determination of the amount of the current overhead cost and how much sales volume is required to cover the overhead.

Table 3.13 provides an example. Assuming an overhead of $369,111 per year, a sales volume of $1,904,831 would represent a business with the required amount of gross profit to permit recovery of overhead and remainder of cost items and to produce a net profit as shown.

TABLE 3.13 Sales Budget Calculation

	Amount	Percent of sales
Material cost	$ 696,778	36.58
Direct labor cost	521,397	27.37
Payroll tax (30% of labor)	157,028	8.24
Miscellaneous direct expense	61,473	3.23
Subcontract expense	59,799	3.14
Prime cost	$1,496,475	78.56
Overhead	369,111	19.38
Profit	39,242	2.06
Sales	$1,904,828	100.00
Minimum working capital	$ 190,828	Sales/10
Net worth	247,829	Working capital/0.77
Allowable building expense	$ 27,261	11% of net worth
Allowable truck expense	34,696	14% of net worth
Allowable tool inventory	12,391	5% of net worth
Allowable furniture and fixtures	24,782	10% of net worth
Total fixed assets	$ 99,130	40% of net worth
Inventory allowance	24,782	10% of net worth

Man-hours to be worked @ $14.00 per hour = $521,397/$14 per hour = 37,243

Number of men working an average of 50 weeks (2,000 hours) = 37,243/2,000 = 19

TABLE 3.14 Application of Overhead on Labor and Material

Labor Rate When All Overhead Is Applied on Labor		
Labor Burden:		
Overhead cost/man-hours = $369,111/37,243		$9.91 per man-hour
Labor rate:		
Labor cost	$14.00	
Payroll tax @ 30%	4.20	
Prime cost	$18.20	
Overhead	9.91	
Profit	1.05	
Total	$29.16	$29.50 per man-hour
Labor Rate When 25% Is Added to Material Cost to Defray Part of Overhead		
Budgeted material cost	$696,778	
Calculated: 25% of material cost	174,194	
Budgeted overhead cost	$369,111	(100% of overhead)
Less: 25% added to material	174,194	(47% of overhead)
Overhead to add to labor	194,917	(53% of overhead)
Labor rate:		
Labor cost	$14.00	
Payroll tax @ 30%	4.20	
Prime cost	$18.20	
Overhead	5.23	($194,917/37,243)
Profit	1.05	
Total	$23.48	$24.50 per man-hour

Overhead added to estimated prime cost: $\frac{369,111}{1,496,475} = 25\%$

Profit added to estimated prime cost: $\frac{39,242}{1,496,475} = 2.62\%$

Table 3.14 illustrates the result of applying 25 percent of overhead to material.

The budget is an intelligent projection of anticipated conditions during an ensuing year; if the conditions fall short of the projections, it will first be necessary to cut the overhead costs and then to reevaluate the budget.

If there are three different electrical departments involved in the business which are constituted differently, then management would proceed with the formulation of this budget for each of them in the same manner as for the single department except that budgeting for each department would involve proration of overhead expenses according to its actual needs. For example, consider three departments with vastly different labor burdens and overall makeup:

Department	Man-hours	Labor burden
Department 1—construction	240,000	$1.66
Department 2—service	52,500	$3.51
Department 3—line	210,000	$0.91

Budgeting for these departments would be as shown in Table 3.15. Notice the disparity in all the elements of cost between each of the departments shown in Table 3.15. It should be noted that the line department furnished very little material for their work, but rather installed material furnished by others. It is important to understand that it would be a serious misrepresentation to lump these figures together to show an overall company average of total costs for each category of material, labor, and so forth; such a figure would be meaningless from a statistical standpoint. (This example points up an important fact: that it is necessary to know the background of any statistics before comparing them with your own. When, and only when, they are comparable they provide useful guidelines.)

It is also necessary to budget each item of overhead cost in accordance with Table 3.7, "Summary of Overhead Expenses," where average costs for the industry are shown for businesses of different sizes. It is this budget that determines the volume of sales required to make a profit, which is the all-consuming objective of every business.

The budget is of prime importance in financial management because it provides the basis for determining pricing policy and sales goals for the current business period. Once it has been determined what volume of sales is required to cover the overhead, the next step is to determine how the sales will be made and by whom. Table 3.16 shows the way in which each individual salesman is allocated a quota of the man-hours of work upon which the budget is based.

Preparing a plan for the coordination of resources and expenditures is a primary management responsibility, but developing a plan for control as the year progresses is of equal importance, and provides the basis for motivation to equal or surpass the goals.

Financing Sources for Electrical Contractors

Five types of monetary backup are or may be required in the electrical contracting business:

1. Permanent working capital which is manifest in the net worth

TABLE 3.15 Budget for Fiscal Year 19XX: Summary of All Departments

Cost	Department 1: construction		Department 2: service		Department 3: line		Total	
	Budgeted amount	%	Budgeted amount	%	Budgeted amount	%	Budgeted amount	%
Material	$2,082,200	48.2	$269,700	31.0	$ 54,150	3.0	$2,406,050	34 4
Labor	1,382,400	32.0	278,400	32.0	911,520	50.5	2,572,320	36.7
Direct job expense	358,600	8.3	43,500	5.0	144,400	8.0	546,500	7.8
Equipment	13,200	.3	25,800	3.2	347,540	19.2	386,540	5.5
Commissions	12,960	.3	17,400	2.0	57,760	3.2	88,120	1.3
Total direct costs	$3,849,360	89.1	$634,800	73.2	$1,515,370	83.9	$5,999,530	85.7
Overhead	397,530	9.2	184,220	21.1	190,430	10.5	772,180	11.0
Profit	73,110	1.7	50,980	5.7	99,200	5.6	223,290	3.3
Total sales	$4,320,000	100	$870,000	100	$1,805,000	100	$6,995,000	100
Man-hours	240,000		52,500		210,000		502,500	
Labor burden	$1.66 per man-hour		$3.51 per man-hour		$0.91 per man-hour			

TABLE 3.16 Budget of Total Sales for All Departments

Department 1			Man-Hours	Sales Budgeted
Salesmen's quotas:				
Bud Quinn			80,000	$1,440,000
Dick Thompson			15,000	270,000
John Hallison			22,000	396,000
Dwight Johnson			93,000	$1,674,000
			210,000	$3,780,000
Plus: Change Orders			30,000	540,000
Total man-hours			240,000 @ 1.66	
Total sales				$4,320,000
	Man-Hours			
Department 2	Residential	Commercial	Industrial	Sales Budgeted
Salesmen's quotas:				
Bob Lindgren	6,000	3,000	3,500	$ 207,200
John Hallinan		8,000		132,560
Doc Easley		1,000	10,000	182,270
Dick Thompson	1,500	10,000	1,000	207,125
	7,500	22,000	14,500	$ 729,155
Plus: Service department	2,000	4,000	2,500	140,845
Total man-hours	9,500 @ 3.80	26,000 @ 2.15	17,000 @ 1.85	
Total sales				$ 870,000
Department 3			Man-hours	Sales Budgeted
Salesmen's quotas:				
Dick Thompson			50,000	$ 430,000
Doc Easley			120,000	1,032,000
Dwight Johnson			40,000	343,000
Total hours and sales			210,000 @ 2.25	$1,805,000
Total sales all departments				$6,995,000

2. Supplementary working capital which is supplied by the sale of debentures
3. Temporary working capital which is borrowed from a financial institution to finance peak business periods
4. Loans on inventory and real estate
5. Leasing of trucks and equipment instead of ownership

Permanent working capital

The source of the cash acquired to organize a business determines the structure of the business, as follows:

1. If an individual can swing it alone, he may have private ownership or he may incorporate and pay income taxes as an individual.
2. If he does not have enough as an individual, he may take on a partner and operate as a partnership or as a corporation with the option of paying income taxes as individuals.
3. Several individuals may chip in and form a corporation by buying stock from the corporation. They may also pay income tax as individuals if 10 or fewer stockholders are involved.
4. A public stock sale may be made and corporate procedures will apply to all affairs.

Supplementary working capital

The sale of 5-year debentures to an underwriting firm, mortgage house, or insurance company affords long-term working capital which can greatly improve the financial structure of the company and broaden the capital base. These promissory notes generally demand a high rate of interest, usually accompanied by a share of the profits and subordination of all other debt to them.

Temporary working capital

There are three principal ways to borrow temporary working capital from a bank:

1. *Open line of credit:* This highly desirable form of low-interest borrowing is available in a maximum amount equal to the net worth of the company. Usually the amount of the credit line is established annually. This form of short-term bank loan is unsecured, and must be retired completely for 60 days every 2 years.

2. *Secured line of credit:* This form of bank credit is secured for short-term loans by stocks, bonds, or cash.

3. *Accounts receivable financing:* This form of bank financing can be very important to a contractor who has successfully bid more work than he could normally handle. It is embarrassing to have to pledge individual accounts and have the customer duly notified; however, many states have legislation which makes it possible for the bank to receive an overall pledge of accounts without customer notification. Complete details of the accounts are required on a running basis by the bank, and the rate of interest is usually quite high. Retainage on the receivables is not accepted as collateral for the loans.

Loans on inventory

Besides paying 1 percent per month or more interest, it is necessary to take an inventory and keep daily records of receipts and disbursements in order to borrow on inventory. Loans up to 80 percent of the cost of inventory are available, and keeping values under this poses a real problem if the warehouse is very active.

Real estate

Pledging real estate as collateral for a business loan is a very desirable means of borrowing capital. Such a loan can be set up for 15 or 20 years if desired, and provides stability to the capital structure.

Leasing

Many firms lease trucks, autos, or office furniture, often offering lease agreements with attractive buyout features. This appears to be a highly desirable method of retaining working capital in the business, while enjoying the benefits of equipment which lightens the workload and reduces cost.

Chapter

4

Effective Marketing

Marketing Concepts

Marketing is the process of impressing on the minds of potential customers the company's ability to provide a service that meets their needs. Effective marketing involves the following coordinated activities:

1. Determining and influencing customers' needs
2. Providing a competitive product which is adequate and acceptable
3. Placing the company's services at the disposal of customers and providing prompt and efficient responses to their requests
4. Creating the proper image of the company through company actions as well as advertising and promotion
5. Establishing sales objectives and creating plans to achieve them
6. Creating an acceptable pricing policy which is competitive but adequate to provide a profit

Electrical contractors are in the business of providing a service which makes it possible for customers to use electricity. While power companies are in the business of generating electricity, contractors bring it into the facility to be used for lighting and other purposes. Very often the process of bringing electricity into a facility is done in accordance with plans prepared by a design engineer; however, many times it is necessary for the contractor to provide the planning required to utilize the electricity in accordance with the customer's needs.

Although much of the contractor's sales effort is directed toward competitive bidding, since being the low bidder is the predominant factor in making the sale, it is also necessary and desirable to expend some sales effort in the areas of direct contact with the customer and the implementation of advertising and sales promotion.

It is, therefore, necessary to develop "sales-minded" personnel in all divisions of the company. It is also necessary to coordinate all the interdepartmental activities, with the ultimate goal being to meet customers' needs in a superior manner in the hope of developing a relationship which leads to repeat business.

Since competing contractors draw their workmen from a common labor pool, it is superb operating performance that distinguishes the services of one contractor from those of another. All employees must have a customer-oriented attitude toward providing service, and the company must focus on motivating, understanding, and reaching customers in an acceptable manner. It is essential for the electrical contractor's overall marketing and sales program to (1) create an image of service and (2) develop a volume of business that is sufficient to provide good, affordable service to customers in a profitable manner.

A budget should be prepared annually for each salesman in each category of sales activity, and market surveys should be performed on a continuing basis to determine the volume of sales potential and the portion of that volume which the company seeks to attract.

Since planning and scheduling are essential to good performance in this business, it is necessary for the electrical contractor's representatives to sell service on the basis of proper coordination of the job with the availability of electrical equipment as needed during the progress of construction. It is essential for the sales personnel to execute detailed paperwork in a precise and timely manner and to communicate accurately to the construction department and the service department the requirements of the job in order to facilitate proper performance of the work and good customer relations.

The electrical contracting business is somewhat seasonal; therefore, it is necessary to plan and to mount an extensive sales effort aimed at scheduling work for the off-peak seasons. It is important for the sales personnel to recognize peak loads of estimating and processing, and in some cases it is necessary to coordinate the limitations imposed in these areas with the customer's requirements.

It is incumbent on the marketing organization to be intimately familiar with the latest products and techniques of installation. Ongoing training programs are required to ensure continued excellent performance.

Potential Market for Electrical Work

It is difficult for a contracting business to operate successfully and profitably unless it is located in a business area where there is an adequate market, i.e., a market with potential for sufficient sales to make a profit.

Statistics indicate that the total volume of electrical contracting work in America is enormous, about $35 billion for the year 1989, but this market is distributed throughout the country in a manner which may not necessarily provide a suitable local market. There is a great market somewhere in this country, but a *good* market is mandatory for success. This underscores the importance of conducting thorough market surveys to make certain that a market large enough to support the business exists.

There are approximately 45,000 electrical contractors in the country, of which about 28,300 have 1 to 14 journeymen and do approximately 9 percent of the work; 11,300 contractors have 5 to 20 journeymen and do about 18 percent of the work. And 5,400 contractors who have more than 20 journeymen do approximately 73 percent of the work. An average total of 422,500 journeymen are employed by the industry during a normal year. The International Brotherhood of Electrical Workers (IBEW) claims a membership of 185,000 workers.

Sources of Business

It is unrealistic to envision salesmen going from door to door to solicit business for an electrical contractor; rather, this type of specialty sales must be pinpointed to sources which provide repetitive business, such as general contractors, construction managers, and purchasing agents or operating engineers for larger industrials and institutions.

In addition to a salesman having a regular, authentic calling program it is essential to follow sales leads such as provided in the *Daily Construction News Services* and the *Engineering News Record,* where jobs in all stages of development are announced for follow-up.

There are several clipping services available which provide newspaper and magazine clippings for any area or all areas in the country. These services forward to the subscriber clippings pertaining to specified sizes of jobs which are proposed for specific areas of the country in which the subscriber operates. The following are representative clipping services:

1. Empire State Press Clipping Service, Scarsdale, New York
2. Allen's Clipping Service, San Francisco, California
3. Luce Clipping Service, Salt Lake City, Utah

A marketing manager should maintain a file of local newspaper and magazine clippings, as well as announcements of proposed construction jobs and similar activities, by subscribing to one of these services.

Building a Company Image

It has been said that "a Company is the lengthened shadow of a single individual." This expression is probably true of all aspects of a business but is especially true of marketing.

To be successful, the company must be known in the community or market area for the services it performs. There are many ways this is accomplished, but the following are the most productive for the contractor himself:

1. Membership in a service club like Rotary or Optimist
2. Membership in a country club
3. Activity in an industry association
4. Membership in a Chamber of Commerce
5. Activity in a political party
6. Activity in community affairs such as the city council or school board
7. Activity in cultural affairs such as symphony concerts or the ballet
8. Membership in a religious institution of your choice

In conjunction with participation in these activities, it is essential to let people know in a gracious way what your business is, and what you might do to serve them. At the same time it is important to look successful and put up a good front.

Advertising and promotion are also important tools to use to enhance the company's image.

Organization of the Marketing Department

When considering the organization of the marketing department it is necessary to recognize what functions are performed by that division of the business; these are shown in the chart in Fig. 4.1. The number of persons employed to perform these functions must be related to the size of the business and its capabilities; the business must be able to provide services of the type being solicited.

In addition to providing enough sales to meet the requirements of the budget, the marketing department must place the company's services at the disposal of its customers and provide prompt and efficient responses to their requests; therefore, the company must be organized and prepared to perform the services before solicitations are made to customers.

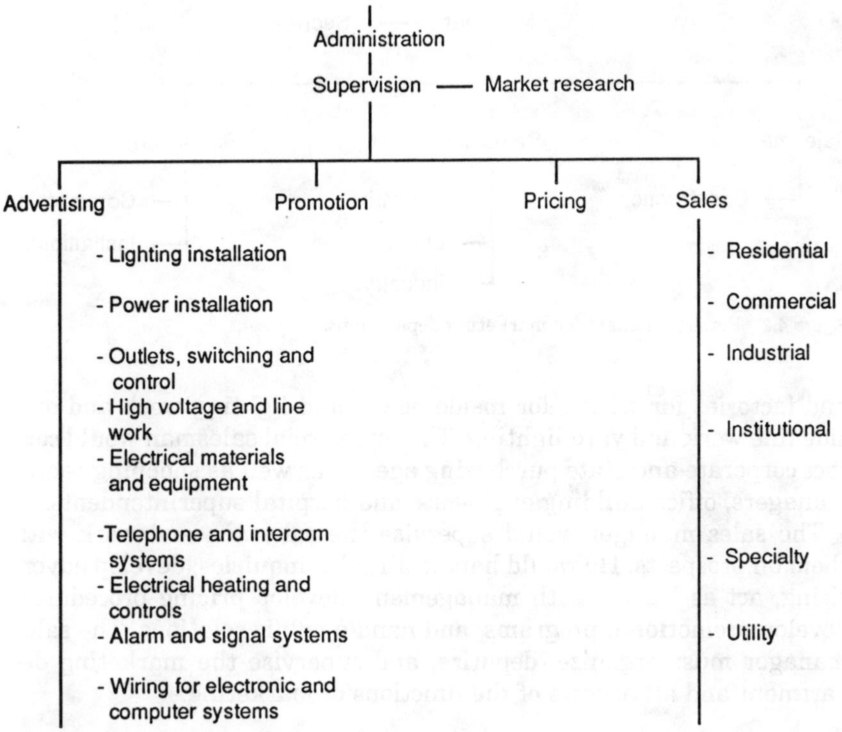

Figure 4.1 Organization of the marketing department.

For example, if the sales department created a customer who could provide 50 service orders per day of 3 hours duration each, the company would have to have 17 service trucks available to serve them as well as all the support facilities required for dispatching, truck repair, materials and tools, warehouse facilities, and billing. Likewise, there isn't any use advertising that you can do line work unless you have the linemen, trucks, tools, and materials that enable the company to provide that type of service.

For a business which has a large number of service trucks, a line department, and a construction department, the personnel chart for marketing would be similar to that shown in Fig. 4.2.

The salesman for construction would develop a liaison with construction management people, engineers, and general contractors. From these contacts, he would solicit jobs to bid and negotiate; he would shag plans and specifications, and he would provide bid proposals and follow up contractor developments. The industrial salesman would contact purchasing agents and operating engineers for plants

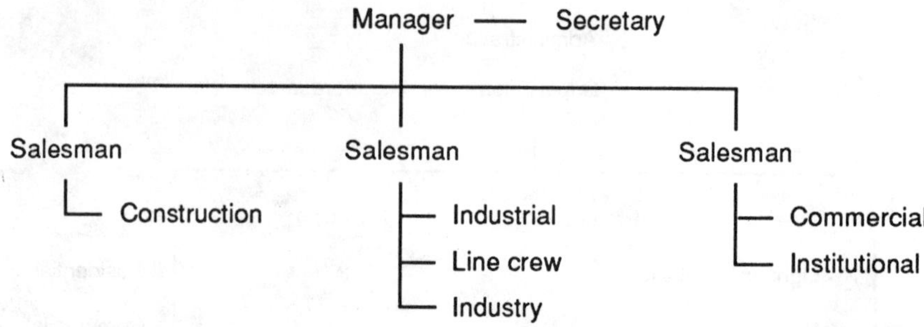

Figure 4.2 Personnel chart for marketing department.

and factories for wiring for inside power and lighting work and outside line work and yard lighting. The commercial salesman would contact corporate and state purchasing agents, as well as shopping center managers, office building engineers, and hospital superintendents.

The sales manager would supervise the salesmen and work with them on prospects. He would handle all sales inquiries, develop advertising, act as liaison with management, develop pricing procedures, develop promotional programs, and handle public relations. The sales manager must organize, deputize, and supervise the marketing department and all aspects of the functions of marketing.

Divisions of Marketing

Almost everyone uses electricity, and most of us depend on it for our very existence. Thus it would appear that nearly everyone is a potential customer for an electrical contractor, and to some extent this is true; more realistically, however, the electrical contractor must be organized properly to provide the various services that customers require. For example, it is very difficult to do a small-job installation without a service department that is organized and equipped to perform that type of work. It is difficult for a contractor who does not have a motor shop to provide service for an industrial plant because many of an industrial plant's needs involve servicing all kinds of motors and controls. Therefore, when determining the divisions of the market in which to sell services, the electrical contractor must first establish the types of work that the business is capable of performing efficiently and effectively.

Essentially there are seven major divisions of marketing activity in which an electrical contractor may become involved: commercial, construction, industrial, institutional, residential, specialty service, and utility. To this could be added specialized installations in a motor repair shop, ceiling installation and interior finishing, and the design and assembly of distribution panelboards and control boards. This chapter will deal only with those areas of marketing in which a wiring contractor may become involved.

Commercial installations and remodeling

Commercial installations involve mainly lighting, power, communications, alarm systems, and convenience and control devices. Typical facilities involved in this class of marketing include automobile dealers, bakeries, banks, grocery stores, department stores, hospitals, hotels and motels, office buildings, printers, restaurants, and shopping centers.

The jobs involved in this type of work are of relatively short duration, usually do not pay large dollar amounts, and require less technical expertise than new construction or industrial installations. However, this category of work has a tremendous volume potential and deserves the attention of the electrical contractor who is well established and is seeking a volume of business which will provide stability for the company. To do this work, the business must have a good service department and a substantial small-contract job department, complete with sales estimators, other estimating capabilities, and job-processing facilities.

The types of marketing involved are principally direct mail and frequent advertising and promotion activities. Sales estimators are particularly essential because the customer relies on the contractor to plan the work that needs to be done and to provide an estimate of cost. The direct sales effort should concentrate on the following:

1. Purchasing agents for large corporations
2. Property agents for realtors
3. Real estate managers for large chain groceries
4. Operating engineers for hospitals, hotels, office buildings, and shopping centers

Construction

Work in this category involves new building construction. Potential customers include architects and architectural firms, electrical engineers and engineering firms in general, general contractors and heavy and highway construction firms, interior decorators, realtors and property managers, the U.S. Army Corps of Engineers, the U.S. General Services Administration, and city and state offices. The electrical work involved is of an almost endless variety, including work on bridges and highways and in office buildings, store buildings, power plants, tunnels, steel mills, cement plants, all types of processing plants, oil refineries, milling plants, and airports.

The jobs involved in this type of construction are generally of long duration, involve large dollar amounts with high retainage, and require a very substantial organization to back up the field forces. This type of work requires a contractor organization with substantial financial integrity, excellent supervision, and considerable business controls to meet deadlines for construction completion dates and to make a profit in the operation of business.

Most new construction work is bid work, and the electrical contractor is a subcontractor; the job is awarded to the low bidder among the subcontractors on a lump sum basis by the general contractor or the construction manager. A contractor's sales efforts in this area are rewarded by being given a preferential chance to bid and often an opportunity to negotiate the bid price with design changes.

In addition to direct sales effort, specialized advertising and some promotions are effective. For example, it is effective to place advertisements in the newspaper when a job is completed, showing a photograph of the facility with special emphasis on who did the electrical work and mention of the general contractor, architect, and owner.

An effective promotion program with general contractors and architects is the mailing of a personal gift on their birthdays.

In order to cement good relations, it is essential that a direct sales effort be made on a continuing basis with general contractors and construction managers. The construction salesman will shag plans and specifications, contact engineers and architects for job information, return plans when the estimator is finished with them, present bid proposals, and follow up contract developments. The construction salesman also assists the contract manager, when advantageous, with owner and general contractor relations during the progress of the job.

Industrial service and installation work

This activity involves new construction, remodeling, and service work. The electrical work involved includes inside and outside lighting, high- and low-voltage power, communications, direct material and equipment sales, motor and control sales and service, and wiring for convenience usage and security.

Considerable depth of organization is required to serve industrial facilities. The contractor should have a service department, a motor shop, and a high-voltage line department, in addition to a complete construction department.

Most sizable industrial plants have their own maintenance personnel, but major changes are contracted out. Small plants require the services of an electrical contractor who can take care of all their requirements with dispatch, and that means depth of personnel and well-stocked truck equipment is required.

Industrial plants that are regular customers should be assigned to a sales representative who can look after their needs on a continuing basis and provide technical assistance when required. Marketing requires continuing contact with the plant purchasing agent or operating engineer by the contractor's sales representative, and the use of birthday remembrances as promotional material. Direct mail pieces which emphasize the stature of the contractor together with new product fliers would be appropriate.

Institutional construction and remodeling

Work for institutions involves both new construction and remodeling. The electrical work includes lighting and convenience work, with limited amounts of power and communications. The types of facilities in this category include schools; city, county, and state facilities; colleges and universities; federal facilities; airports; recreation districts; and libraries.

The volume of work available from this source is not as large as that available from others, and this market is best served by a contracting organization which has a competent service department and a small-job construction department.

The types of marketing required are direct mail advertising and promotion; also, a sales estimator who can make on-the-spot estimates and assist the customer in planning must be available. The direct sales effort for this category of work may be combined with the commercial sales activity, and the following contacts may be fruitful:

1. Purchasing agents of the federal government, state, city, and county
2. Purchasing agents for colleges and universities
3. Operating engineers for airports, military posts, and government institutions like prisons and office centers

Residential and apartment construction and remodeling

Residential new construction involves everything from wiring for individual homes to tracts of many homes, apartments, and condominium projects. Handling this type of work is a specialty, and the work is awarded on a very competitive basis. Therefore, the contractor's organization must be geared in every way to install this work and meet this type of competition, and still make a profit. Developers of tract homes usually do their own construction so they are a source of business for the electrical contractor; this work is secured by direct contact.

Apartment new construction requires an electrical contractor of substantial fiscal capability. A job may involve a single building, or there may be several in a cluster. This type of work is very competitive, but the projects are large enough to be very rewarding. The developer usually does his own construction work, so the electrical work is handled as a prime contract, and is awarded by the developer.

The marketing for both residential and apartment-type work is accomplished by direct sales contact, with promotion in the form of birthday remembrances for the developer.

Residential and apartment remodeling is handled on a small-job basis by service men with a truck.

Specialty service work

Most of the work in this category is done on a small-job basis for other contractors who specialize in air conditioning, ceiling installations, heating, kitchen remodeling, and temperature controls. When a job done by one of these contractors is ready for the electrical work, it is essential that this work be done quickly so that the job can be tested and completed. In order to do this type of work as expeditiously as required, the electrical contractor should have a sizable service department which has considerable flexibility and capability to serve a customer on short notice.

The most effective marketing effort consists of direct customer contact made by a sales representative to the specialty contractor.

Utility work

Public utilities require many services from the electrical contractor, and all types of electrical work are involved. Potential customers include power companies, telephone and telegraph companies, the Bureau of Reclamation, and water boards.

Electric utilities have found it profitable to employ contractors to install overhead line work and underground distribution work for subdivisions in the outlying residential areas of a town. This work is done under a unit price contract arrangement for both classes of installation work, and the unit price includes the cost of labor, equipment, and trucks and margins for overhead and profit. The electrical contractor usually provides labor only, and the utility furnishes the materials.

The utilities have found that work done in this manner is less costly to them, and they are not committed to the employment of labor and investment in equipment on a continuing basis. This work cannot be done profitably on a small-volume basis; in fact, it requires a volume of at least 10 crews working on a continuing basis. In order to forestall shutdowns due to equipment failure, the contractor must operate his own garage and should have capable crew foremen for field supervision. Work of this type also involves short-wave radios and well-equipped trucks.

Power plant work, substation work, and the construction of transmission lines are also excellent sources of business for the electrical contractor. All such work requires a substantial, well-equipped business with personnel who are technically capable.

Most of the marketing effort for this type of work involves direct contact with the utility personnel who are responsible for getting the work done, and usually the contractor must have had much experience in handling specialized work. Nearly all such work is done under some type of contractual arrangement.

A great deal of electrical work is done for telephone utilities. It involves both new construction and remodeling of existing terminals. Sometimes a telephone utility will subcontract all its installation work in a new commercial building, including the switch, terminals, pulling of cable, and the installation of the telephones. The installation of microwave relay stations is very often done under a contractual arrangement.

Most telephone installation work is done by a special class of labor other than the inside journeyman wireman. The automotive and equipment requirements for this type of work are not as stringent as those for electric utility work. Most of this work is done under unit price contracts or some other form of contractual arrangement.

Marketing of these services is done by direct contact with the proper personnel in the telephone company or power company.

Marketing Methods

As we said at the beginning of the chapter, *marketing* is the process of convincing potential customers that your company has the capability of doing work which satisfies their needs. It is essential to determine what methods the contractor needs to use to convince the potential customers, and the following methods must all be considered:

1. Direct sales
2. Advertising
3. Promotion

The purpose of these forms of marketing is twofold: to place the company's services at the disposal of the consuming public, and to create the proper image of the company.

Even after the consumer is successfully attracted to the company, it is almost always necessary to bid for the work, and as often as not on a competitive basis.

Direct sales

Direct sales occur when a salesperson contacts a prospective customer in an attempt to create business. This type of effort may take several forms, but the most important is "creative sales" which manifests itself in new business for the company.

Since the clientele of a contractor may be constantly changing, it is mandatory that there be a continuing sales effort geared to creating new business to replace customers who have moved or gone out of business. Direct sales efforts can also focus on working with new businesses which result from population increases and a growing economy.

Advertising

The most effective types of advertising for the electrical contracting business, given the budget normally available, are direct mail and radio.

Direct mail. This is a very effective means of pinpointing advertising to a selective target such as architects, general contractors, or industrials, with topics of special interest. Direct mail must not be of a general nature. It must deal with subjects of a specialized nature, and the mailing list must be accurate and kept up to date. Possible specialized subjects are (1) yard lighting (protective), (2) office lighting, (3) garden lighting, (4) alarm and burglar systems, and (5) electric heating.

City newspaper. The best use of city newspaper space is to publicize new jobs obtained or sizable jobs completed. This also provides an opportunity to recognize the architect and general contractor, which improves your relationship with good sources of business.

Radio. An effective use of radio is the sponsorship of a weathercast at a time of day when it is of particular interest to the public, such as at breakfast time or in the morning when businessmen are driving to work.

Neighborhood publications. Advertising space in neighborhood publications is less costly than space in city newspapers. They can be good places to advertise for service work and they will sometimes print public relations articles.

Advertising budget. Every business must spend some money for advertising and promotion, but the question always is: "How much?" For guidance we must turn to NECA statistics for each of the four groups of businesses, as follows:

Sales volume per year, thousands of dollars	Percent spent for advertising and promotion	Total outlay, dollars
712	0.41	2,919.20
1,477	.23	3,397.10
2,830	.27	7,641.00
6,042	.23	13,896.00

These statistics are industry averages; actual expenditures vary widely.

The results of these expenditures are very seldom felt or realized by the cash register, but the results are real, and the expenditures are essential.

Promotion

Promotion involves the use of handouts to make a lasting impression on individuals. These handouts involve such things as pocketknives, pencils or pens, desk sets, and thermometers, all inscribed with the name of the company and a phone number.

Promotion may also involve the support of an athletic team such as a softball, baseball, or basketball team. It might involve the support of an automobile racer or a speedboat or any kind of a special event of prominence. These activities place the name of the company in front of the public in a prominent manner and make a good and lasting impression.

Pinpointing the Market

Direct sales, advertising, and promotion all have a definite place in the electrical contracting business and the successful contractor will find it necessary to use all of them. The purpose of all three methods is to influence someone to use your company for their electrical work. However, the spending of dollars for these items must be carefully planned and in no case should these dollars be expended in a "shotgun" manner which will produce little if any return on your investment.

The terms *demographic data* and *psychographic data* are encountered regularly in any marketing plan and apply to the business, industrial, and professional markets just as they do to the consumer market.

To pinpoint the electrical market in your area you first must identify the demographic data of the business, industrial, and professional customers and prospects. This data would be:

1. Address or addresses
2. Volume of business (growing, static, decreasing)
3. Position in business, industrial, or professional field

Next it is necessary to develop the psychographic data, which would consist of:

1. The purchasing picture
   a. Who makes the buying decisions? You must learn: name or names; position or positions.
   b. Are buying decisions made at any particular time? When?
   c. What is the decisionmaker buying now? Your products or services? Competitive products or services? *Why?*
   d. Is there potential for additional business with your present customers? What is the potential? Greater quantities of products or services you are now selling? Additional products or services?
2. The purchasing influences
   a. Current users?
   b. Company representatives? Sales personnel? Service personnel? Jobbers? Distributors?
   c. Product features?
   d. Services?
   e. Advertising? Whose? In what media? How often? With what message?

Sources for data needed

Sources for the needed data include:

- Your own organization
  - Managerial personnel
  - Sales personnel
- Communications media
- Industry or business or professional groups
- Industrial, business, or professional publications
  - Their articles and editorials
  - Their principals
  - Their sales representatives

House organs

Annual reports

Financial analysts

Consulting firms

Once you have determined what market for your services is available in your desired market area, and which prospective customers you want to develop, you should then examine the competition for that market. You will need to *know your competition:*

1. Identity and market strengths
   a. Name
   b. Number of employees
   c. Types of work or services competing with your company.
      (1) Strengths
      (2) Shortcomings
      (3) Volume of business
      (4) Share of the market
2. Sales efforts
   a. Who are their sales personnel?
   b. What are their sales policies and practices?
   c. How do they promote their business? Pricing? Discounts? Any other specialty items?
3. Advertising methods
   a. How much do they do?
   b. What is their advertising coverage?
   c. What media do they use? Broadcast (radio and television)? Print media (trade journals, magazines, newspapers)?
   d. Direct mail
4. Advertising approaches
   a. What is their strategy?
   b. What is the sales appeal? Do they have a central claim that they make (are they better, cheaper, larger)? What message do they use to support their claims?
   c. Do they have any creative techniques which you could use?

Some sources of information about your competitors. Sources for the information you need to gather about your competitors include:

Industry associations and conventions

Company news releases and financial reports

Business magazines, local publications, business sections of local newspapers

The Small Business Administration

Your suppliers

Your own employees (e.g., sales and service people)

Monitoring of the competition's advertising

Advertising and promotional budgets for most electrical contractors are rather small when compared to the several million dollar budgets of large corporations. Therefore, every dollar you spend on advertising and promotion must do double duty and should be made to build your company name and build your business.

Measuring Advertising and Promotional Efforts

There are five basic requirements against which all your advertising and promotional efforts should be measured. These are:

1. *Circulation:* How broad a base is served by the medium or method you choose?
2. *Usage:* How many businesses or individuals use the medium or product you use for your advertising or promotion?
3. *Retention:* How long will your message be retained for a repeated reminder of your company?
4. *Economy:* Aside from initial cost, what is the ultimate return on your investment? What is the cost for each contact made?
5. *Results:* How much direct action can possibly be achieved by your advertising or promotion?

Creative advertising of professional quality will stimulate prospective customers' interest in your company. You should choose your advertising agency carefully, and before assigning them the job of producing your brochure or ad, you should examine many samples of their previous work. Directional advertising will direct prospective customers to your place of business when they need your type of services.

There are some time-proven principles for creating advertising copy that will attract prospective customers, and there are other choices that should be avoided. First the things that should be avoided:

1. White letters on a black background (this is known as the "bargain basement look").
2. A mix of too many lettering styles and/or graphic devices. Do not

combine italics and block-style lettering; this will only detract from the pleasing and attention-getting quality of your ad.

3. Avoid the "circus poster" look.
4. Do not try to tell too much or make too many sales points as this will clutter your ad and cause the reader to lose interest.
5. Do not crowd the space—this is known as the "pitchman pitch."
6. Leave plenty of room for your name to be displayed prominently along with your address.

Some characteristics which will improve the impact of your ad are:

1. Use dark letters on a light background. (Research has proved that this is the easiest to read.)
2. Keep graphics simple, letter styles limited and harmonious.
3. Select and mention only a few product or service benefits and avoid clutter. The ad copy must be clearly written.
4. Open up your ad. White space invites looking and reading.
5. Use large type for your name and address.

Sources of Customers

There are many sources of customers for the electrical contractor and some of them are:

1. Those customers recommended to you by other satisfied customers
2. Those who have been influenced by your advertising.
3. Those who have been solicited by your sales force
4. Passers-by who saw your place of business or your vehicles
5. Former customers who previously did business with you
6. Newcomers to the area who are looking for a firm with which to do business
7. Customers who have an emergency and need help now
8. New customers who are dissatisfied with the product or service of your competition
9. Those who have a need for your services who have not required electrical service before
10. Those who are comparing price, quality, and service or looking for estimates or bids

To keep the customers you have you need to:

1. Be reliable at all times; establish a record (years in business, size of business); handle nationally recognized products
2. Handle only authorized products that fit the job you are doing
3. Develop special features for your business and unique points that separate you from the competition
4. Offer a complete service that will enable you to complete the job in a professional manner and assure that the job is not only completed on time but that it is right the first time
5. Have attractive slogans, logos, brochures, etc.
6. Have a location convenient to the area served by your company and your customers

Brochure

Many electrical contractors have discovered that an attractive brochure can be a "door opener" when approaching a new general contractor, building superintendent, architect, or engineer. The beautifully designed and printed brochure probably has more retention value than any other printed form of advertising or promotional material. In addition the brochure affords you the opportunity to present your company history and capabilities in greater detail than other types of advertising materials.

Because most electrical contractors have limited advertising and promotional budgets, the cost of producing the brochure is an important consideration, and the key to better production and lower cost is planning. A total of at least 120 days must be allowed from the start of creative work to delivery of the finished product from the printer. The best-looking brochures are the result of careful planning of the layout, the copywriting, and the choice of photographs. Overbudget costs can be avoided by allowing plenty of time for each phase: photography, retouching, typesetting, proofreading, engraving, and printing.

There are many ways to save money on the production of a brochure. Some of them are listed below.

1. Use only the experts. It pays to use professionals in this endeavor. A good art director can save you money and give you a better product.

2. Get three printing bids. Be sure the printer has the right equipment for your job. In order to get printing bids you must have a set of

specifications that will tell the printer the size of your brochure, number of pages, weight and type of paper stock, how many colors will be used, and the number of color photographs that will be included.

3. Put the production schedule in writing. Allow at least 16 weeks from start of production to delivery. Outline each step in the process and include dates for approval of copy and layout, selection of photographs, preparation of mechanicals, delivery to printer, first proof from printer, press approvals, and distribution. Do not assume that every step will be right the first time—allow for corrections.

4. Have the printer order the paper early. The cost of paper continues to go up and has become one of the major cost factors in producing a brochure.

5. Carefully edit and approve copy before having the type set. Resetting type is a costly, unnecessary waste of money. The time for changes is while the copy is still in manuscript, not after it has been set in type.

6. Make changes on the mechanical copy, not the engraved plate. The mechanical is a good representation of what the printed page will look like. There is usually a tissue overlay on the mechanical, and you should use the tissue to make any corrections.

7. Use transparencies for color photographs and you will get higher-quality reproductions than from a color print. If you are producing a black and white brochure, a black and white print is the preferable medium.

8. Use fewer large photographs instead of a large number of small ones. Each color photograph must be "separated," so the more you use the more you pay.

9. Special effects in your brochure will be expensive so if you plan on using some be sure to include the cost in your budget.

10. Consider the cost of mailing or delivering your brochure. Do you wish to mail it first class? Would you rather have a larger, heavier brochure even though you would have to mail it third class?

Forms of Bidding for Electrical Work

Most of the jobs secured by an electrical contractor require some form of contractual arrangement which involves a price for the work. The form of bidding is generally determined by the urgency with which the job must be completed, and the completeness of the plans at the time of bidding. The form of bidding is determined by the awarding authority, but very often it can be changed for the owner's benefit by negotiations between the awarding authority and the electrical subcontractor.

The following are the forms of bidding most frequently used:

1. *Lump sum:* Total price includes the cost of material, labor, direct job expense, prorated share of overhead cost, and profit.
2. *Unit price:* Prices are submitted complete with labor, job expense, overhead, and profit on units of material—for example, foot of conduit including fittings, foot of wire including terminals, outlets complete with support, fixtures including lamps and parts, panelboards, transformers.
3. *Reimbursable cost plus a fixed fee:* All cost of material, labor, and job expense are reimbursed to contractor according to actual cost. Fee for a specific scope of work is estimated to include overhead and profit.
4. *Reimbursable cost plus a percentage:* All costs of material, labor, and job expense are reimbursed to the contractor according to actual cost. Overhead and profit are paid as a percentage of prime cost.
5. *Guaranteed maximum price:* The job is bid on a lump sum basis, but the owner is assured that a reimbursement will be made if savings can be effected at a price less than the estimated bid. Very often the savings is split with the contractor.
6. *Time and material:* Labor is billed at a resale price per hour for foreman, journeyman, and apprentice. Material is billed at a resale market price less a discount. Overhead and profit are included in the labor and material prices. Direct job expense is billed at contractor cost.
7. *Labor only:* Owner furnishes material. Contractor furnishes labor, tools, and equipment. Contractor bids on a lump sum basis including labor cost, payroll tax, overhead (as if contractor were furnishing material also), and profit.

Negotiating Electrical Work

Negotiating is defined as the process of conferring with another person for the purpose of arriving at a settlement of some matter, which in our case is the sale of an electrical construction contract. The negotiating procedure is used instead of outright bidding in order to save time in the construction process. The time that it takes to prepare complete bid documents can be used to good advantage in getting the job under way, and for a competent contractor it is also a good way to save money and bring the job in within the budget.

When conducting such negotiations, the electrical contractor will probably need to design the job before it can be estimated; in order to do this he must be provided with an architectural drawing of the facility and complete information as to the power and lighting requirements. With this information the contractor can prepare an electrical drawing from which an estimate can be prepared and a bid proposal drawn up for submission to the owner or awarding authority. Sometimes the owner will involve more than one contractor, but usually negotiations are carried on with just one.

When the contractor presents his proposal, the plan is reviewed carefully and additions or deletions are made to the plan with subsequent modifications to the price. If the proposal is then within the budgeted allowance of the owner, the work is approved and proceeds. This procedure is generally a last minute transaction because time is of the essence. Once the owner has the funds and has made up his mind to proceed, the cost of delays is very significant.

Sometimes the electrical design is done concurrently with the construction work, in which case it is impossible to present a firm price bid. In this situation the following types of contracts may be negotiated:

1. *Cost plus a fixed fee:* This type of contract involves the electrical contractor submitting an estimated fee that includes overhead and profit; the fee is one that the contractor feels would be adequate to support the cost of material and labor and direct job expenses, which he would bill monthly in accordance with actual cost.
2. *Cost plus a percentage:* This type of contract would require the contractor to determine a percentage which would be added to the actual cost of material, labor, and job expense.
3. *Cost plus with a guaranteed maximum:* This method requires an outright estimate.

Design and Build

When an electrical engineer designs a job for an architect, it is incumbent on him to be extremely thorough in order to avoid the possibility of a malpractice situation. Therefore, a well-informed contractor can probably effect considerable cost savings through redesign and substitution of materials without greatly affecting the quality of the job—

thus helping the owner achieve *his* objectives, which are a good job at the lowest price. At this time in our economy, with the high cost of services, it is mandatory to accomplish the desired ends at absolute minimum cost, and making use of the ingenuity of competent people is one way of doing this. Usually the awarding authority takes competitive bids according to a designed set of plans and specifications and then works with the low bidder to reduce the cost through redesign and substitutions of circuitry, fixtures, and equipment.

Joint Ventures

There is nothing as satisfying as being able to successfully operate a business on an independent basis, but there are some situations where a joint venture is highly beneficial. These are:

1. Your own working capital is inadequate to swing a large job which comes into your working area.
2. Another contractor with whom your have a close working relationship has close connections with a general contractor who is coming into your area with a job under contract; your friend can get the job but does not wish to work alone in a strange jurisdiction, so he offers to join forces with you in a joint venture.
3. Converse situation to item 2.
4. A special type of work makes it desirable to join forces with another contractor who is already familiar with this type of work.

Before teaming up with another contractor in a joint venture it is extremely desirable to know his financial condition and to have an airtight agreement drawn up establishing the basis for the working relationship.

Pricing Policy

To ensure good customer relations, it is essential that every contractor establish a pricing policy. The policy must be geared to establishing a good reputation in the community and creating an image of the company as fair, honest, competitive, and successful in the conduct of its affairs. In all matters of pricing, the objective must be to sell at the lowest price which is compatible with making a profit.

The nature of the electrical contracting business is such that a large volume of sales is required to support an organization large enough to provide the services to which its customers are entitled. While electrical contracting is not necessarily a large business, many important processes are involved in providing electrical installation work to customers in an acceptable manner. Electrical contractors cannot price correctly unless they operate properly. It is necessary to provide service as efficiently as possible at the lowest possible price, always recognizing the need for quality and meeting the requirements of the performance schedule.

The exercise of ingenuity is essential in arriving at a proper price. Specifications govern the limits of conduct, but the use of intelligent substitutions of material and manipulation of overall design and technical features within acceptable parameters are part and parcel of proper pricing.

Change Orders are part of the electrical contractor's budgeted sales, and margins are established according to adders which are acceptable to customers. Because the customers are "captive" once the contractor begins working for them, it is essential that cost records be accurate, that job factors be properly applied, and that margins be added in an acceptable manner.

Price agreements established by a salesman's efforts and accepted by a customer may be "arm's length" transactions, but even then the electrical contractor has the responsibility not to betray the customer's trust and to price the product in an equitable manner.

At all times and under all conditions, electrical contractors must conduct their businesses in a manner acceptable to their customers; they must strive to maintain a reputation of trust, honor, and confidence; and they must serve all their customers' electrical needs on a continuing basis.

Contract Documents

The very term *electrical contracting business* implies that most sales involve some type of contractual arrangement or sales proposal, and in fact it is essential for any building trades contractor to be intimately familiar with the elements of contracts.

The definition of *contract* is "a specific agreement, enforceable at law, between two parties for the execution of a definite exchange of services for a specific amount of remuneration whereby each acquires a right to what is promised by the other." Thus, in the formation of a contract the following elements are present:

1. Competent parties
2. A lawful subject matter
3. A proper consideration
4. A genuine agreement or mutual assent

While a construction contract is a very comprehensive document, there are a few clauses which should always be of particular concern to an electrical subcontractor. Clauses covering the following always deserve attention.

Approvals Proper timing is essential for the submittals for approval and the return to the vendor. Serious interruptions to the progress of the work can occur if this process does not "flow" properly.

Changes It is essential to identify those authorized to approve Change Orders, and no work should be done on Change Orders without written approval. Also, an extension of time may be involved.

Claims Most claims are against the owner rather than the general contractor, but the subcontractor has a contract with the general contractor only; therefore, the subcontractor must be entitled to all rights, remedies, and redress against the general contractor which the general contractor has against the owner.

Concealed Conditions The requirements of the plans and specifications are all inclusive and complement each other. Requirements may be spelled out in the specifications but not shown on the plans.

Contract Administration It is important for a single person to be designated as administrator of the contract.

Damages Under no circumstances should the subcontractor be liable to the general contractor or owner for any losses or damages in the nature of partial or complete loss of use of the project; loss of interest, revenue, or anticipated profits; or any loss of a similar nature.

Design Errors How design errors may be reconciled is important in terms of time and money. Certain errors in the wrong places may affect the entire schedule.

Disputes It is essential for the subcontractor to have the same rights regarding disputes as the general contractor has against the owner. If arbitration is specified in the prime contract it must also be specified in the sub's contract.

Indemnity No one should be relieved contractually of liability for their own negligence, nor should a contractor be required to indemnify the owner for damages resulting from the project except those for which the contractor is directly responsible.

Insurance The requirements for the job should be the same as those spelled out in the bidding documents, and the subcontractor should not be required to cover deductibles in the owner's or general contractor's insurance policies.

Interference A means of resolving interference must be provided when two different things are occupying the same space.

Payment (1) The contract must contain a clear statement of entitlement to payment in full upon completion of the work. (2) The contract must specify prompt pass-through for your share of payments received from the owner. (3)

The contract should state that Payment must be made during the month for which the billing was made.

Performance Time You are entitled to receive a complete construction schedule whether or not a penalty is involved, but especially if so.

Retainage Payments involving retainage must reflect only the rate actually being retained by the owner.

Scheduling When the plan of operation specified in the construction schedule is changed, it is very important for the subcontractor to be informed.

Site Inspection All contractors are required to visit the site before bidding, and such a procedure is mandatory whether specified in the contract or not.

Termination of Work The owner must have a means of removing from the job an incompetent contractor or one who is unable to keep up with the schedule, but every effort should be made to modify any clause which provides for strictly unilateral action.

Waivers Many contracts state that the acceptance of the final payment shall be considered as release in full of all claims against the customer. A general contractor's form says final payment constitutes a waiver of all claims by the subcontractor. Care must be taken to insert exception "for those identified in writing which are unpaid."

Chapter

5

Effective Claims Determination

When undesirable and unplanned working conditions which are beyond the control of the contractor occur on a job, and when these conditions cause costs of installation to increase beyond the scope of the contract, measures must be taken to retrieve the costs by developing claims against the owner. In order to develop these claims effectively, it is essential to identify the conditions which caused the increased costs, and then to be able to evaluate them.

In this chapter we consider the various types of claims and the procedures used to evaluate them. Claims fall into categories such as the following:

- Acceleration and decreased productivity in overtime
- Additional engineering costs for extra work
- Extended overhead and job expense
- Financing the job with borrowed capital
- Joint occupancy with construction forces other than your own
- Liquidated damages and time extension

Either individually or collectively these result in greatly increased costs to the contractor. It is essential that proper recognition be given to each of them if they occur during the progress of the job in order to properly document them, and this means that the contractor and his organization must be familiar with each of them and know what to record and how to record it as the work is done.

Acceleration and Decreased Productivity in Overtime Work

When an electrical contractor successfully bids on a job where the construction schedule is based on a normal completion time, it is appropriate to assume that the work will be done under normal conditions. If the owner then decides in the course of construction that the completion date must be advanced, the electrical contractor is faced with two alternatives:

1. Increase the number of men working on the job.
2. Work the existing crew overtime to keep up with the increased work schedule.

In either case, the job is subjected to *acceleration* and the contractor is exposed to increased costs for which he is entitled to reimbursement.

In order to justify a claim for the increased cost, it is necessary for the contractor to illustrate the effect the changed condition had on the manpower loading by comparing the manpower chart prepared at the beginning of the job with the manpower chart of the actual work done under the accelerated conditions. Acceleration nearly always results in overtime, and overtime work results in reduced productivity due to fatigue and change of attitude. The extent of loss of efficiency varies according to the length of the work day and the number of days worked during the week, as shown in Table 5.1.

Table 5.2 shows total overtime hours for an accelerated schedule, and calculation of excessive costs on the basis of reduced productivity during overtime. Table 5.3 gives a summary of total costs due to ac-

TABLE 5.1 Table of Overtime Efficiency Rates

Hours worked	How worked	Overtime applicable to:	No. of overtime hours	Efficiency rates, %
Five 10-hour days	On same job	Hours over 8	2	87.5
Five 12-hour days	On same job	Hours over 8	4	75.0
Five 10-hour days	Workers moved to other job after 8 hours	Hours over 8	2	75.0
Five 12-hour days	Workers moved to other job after 8 hours	Hours over 8	4	68.8
Six 8-hour days	On same job	Sat. only	8	87.5
Six 10-hour days	On same job	Total hours	60	84.5
Seven 8-hour days	On same job	Sat. & Sun.	16	77.0
Seven 10-hour days	On same job	Total hours	70	77.8

SOURCE: Efficiency rates from NECA.

TABLE 5.2 Schedule of Excessive Labor Costs Due to Acceleration*

No. of man-weeks worked	Hours worked during week	No. overtime man-hours involved per week	Hours subject to loss of efficiency	Efficiency	Productive hours	Nonproductive hours	Cost of nonproductive man-hours @ $15 per man-hour
400	Five 10-hour days	10	4,000	0.875	3,500	500	$ 7,500
100	Six 8-hour days	8	800	0.875	700	100	1,500
100	Six 10-hour days	60	6,000	0.845	5,070	930	13,950
200	Seven 8-hour days	16	3,200	0.770	2,464	730	11,040
100	Seven 10-hour days	70	7,000	0.778	5,446	1,554	23,310
			21,000			3,820	$57,300†

*Basis of typical schedule:
- Man-week assumed (period of time, e.g., 48 h, 50 h, 60 h, etc.).
- Efficiency rates from NECA reports.
- Efficiencies of 8-hour days apply to Saturday and Sunday.
- Efficiencies of 10-hour days over 5 days apply to total hours.
- Efficiencies of 10-hour days on 5-day week apply to hours over 8.

†Total loss is $57,300 plus fringes.

TABLE 5.3 Summary of Costs Due to Acceleration

Total hours worked overtime	21,000
Nonproductive hours due to efficiency loss	3,820
Cost of labor @ $15.00 per hour	$ 57,300
Cost of supervision @ 15% of labor	$ 8,595
Total labor base cost	$ 65,895
Cost of fringe benefits (31% of base cost)	20,427
Cost of small tools and supplies @ 7%	4,612
Total net cost	$ 90,934
Job overhead allocation cost @ 15%	13,640
Home office allocation cost @ 4.5%	4,092
Total prime cost	$108,666
Profit @ 10%	10,866
Total billing for acceleration:	$119,532

celeration based on the overtime hours and lost productivity costs in Table 5.2.

Examination of the manpower charts superimposed on each other for the two periods of time will provide information required to evaluate manpower overloading and its effect on productivity.

The following factors are influenced by increasing the number of men on the job when additional workers are needed to do the work that a smaller number would have done under normal working conditions:

1. There is increased supervision cost where, under the working agreement, the amount of supervision is determined by the number of men working.
2. There is increased cost when two men are used to perform work normally done by one man.
3. There is increased nonproductive time resulting from conditions that cause fatigue and change of attitude on the part of workers.

Under normal conditions time spent in the field is allocated, on average, as shown below:

Type of task	Percent of total field time
1. Studying plans and specifications	5
2. Ordering material and tools	2
3. Receiving and storing material and tools	4
4. Moving material to the point of usage	6
5. Tooling up	6
6. Measuring and laying out the work	4
7. Installing	63
8. Performing nonproductive tasks	10
Total	100

When two men perform a one-man task, the above percentages are affected due to overloading, as follows:

Type of task	Percent increase
Moving material to point of usage	2
Tooling up	4
Measuring	4
Installing	30
Nonproductive tasks	9
Total increase	49

Thus the labor cost is increased 49 percent by using two men to perform a one-man job.

Changes Resulting in Additional Engineering Cost

The cost of engineering work directly connected with each change should be claimed separately. In the case of settled modifications where additional direct engineering is claimed and paid in full, such payment is credited to the claim. However, in addition to the direct engineering effort, labor, and material that can be isolated and attributed to a given change, other factors directly and indirectly affect both total costs and the time required for performance, including the following:

- The adverse effect of changes on the overall labor and engineering effort in terms of disruptions in previously planned work schedules, the necessity to hold up phases of the job pending approval of a change, and the attitude of workers on the job
- The cost of estimating the price of the changed work and the time required to perform it, including time spent searching through previous modifications to make sure that the latest contract requirements are considered
- The cost of travel and sometimes lengthy negotiations covering a multitude of separate items within a given modification
- The cost of extra layout and preparation of working drawings
- The cost of additional material approval submissions
- The cost of extra clerical work for purchasing additional materials, maintaining cost records, amending subcontracts, and billing for the changes
- The cost of additional financing required to do more work within a given time

Claims for additional engineering hours are based on the concept that, had the contract drawings and specifications been adequate, it would have been possible to greatly reduce the engineering effort at about the time the construction work was getting fully under way, while instead it was necessary to continue a heavy engineering effort right up to the very end of construction.

For the purposes of a claim, *impact costs* are defined as increased costs that are attributable to the cumulative effect of numerous changes in the original contract, and to the resultant disruptions and inefficiencies introduced into a contractor's construction operations. Impact costs result from factors such as mobilization, demobilization, time spent waiting for problems to be resolved, interruptions to overall planning and scheduling, and inefficiency due to lack of continuity of the work.

In the field, two forms should be used to record engineering deficiencies. These are the Work Stoppage Report and the Engineering Clarification Request. These forms are shown in Figs. 5.1 and 5.2. In order to produce a Cost Estimate and Price Proposal for impact costs from engineering design changes, the backup data formatted below will be required.

Item no.	Description	Engineering hours	Impact on labor hours
1.	________	________	________
2.	________	________	________
3.	________	________	________
4.	________	________	________
And so forth.			
Total engineering hours			________
Total journeyman hours			________

Table 5.4, the summary of Cost Estimate and Price Proposal, shows how to calculate the additional for engineering and field labor which result from these changes:

Extended Overhead and Job Expense

Because of the vast amount of paperwork and the increased management and supervisory time required when it is necessary to process changes and additional work, to remedy design deficiencies, and to resolve conflicts in the drawings and specifications, as well as to accelerate the construction work to overcome other excusable delays, the electrical contractor is required to expend larger sums than estimated

Work Stoppage Report

CPM ____________
Area ____________
Node ____________

Date Work Stopped ____________ Time ____________

Date Work Resumed ____________ Time ____________

Description of Work Stopped ____________

Reason Work Stopped ____________

Responsibility: ______ Contractor ______ General ______ C. O. E. ____________ (Other)

Signed: ____________ Date of Report: ____________

Figure 5.1 Work Stoppage Report.

Engineering Clarification Request

By: ____________ Title: ____________ Date: ____________

Engineering Clarification is Needed as Follows:

Area: ____________ Location in Area: ____________

Drawing Reference: ____________ Work Order Number: ____________

Remarks: (Indicate All Specific Trouble Areas and Drawing Discrepancies)

Figure 5.2 Engineering Clarification Request.

for direct job expenses and for overhead—not only during the original time specified for construction, but for approximately 9 months thereafter. As stated previously, under normal conditions these direct job expenses and overhead costs would have started tapering off approximately halfway through the construction program. However, under

TABLE 5.4 Estimate and Price Proposal: Summary

Labor:	
Impact on man-hours	________
Work Stoppage Reports	________
Engineering Clarification Requests	________
Total	________
Total man-hour cost: X hours @ $XX.XX per hour	$________
Engineering:	
Man-hours @ $XX.XX per hour	$________
Total engineering and labor	$________
Other direct costs:	
Top-level supervision @ 10% of labor	$________
Small tools and supplies @ 7% of labor	$________
Total direct costs	$________
Overhead:	
Job overhead @ 15% of direct costs	$________
Home office overhead @ 4.5% of direct costs	$________
Total overhead costs	$________
Profit: 10% of gross cost	$________
Total billing	$________

abnormal job conditions these costs continue at a high rate after the completion of the original contract work. It is considered that this extension of direct job expenses and overhead represents a cost for which the electrical contractor is entitled to reimbursement.

In addition to these charges, the direct costs for changes in the physical work, acceleration, joint occupancy, and additional engineering work will be claimed separately. These costs are not a part of the extended job expense and overhead claim. Where direct job expense and overhead were allowed in modifications, such allowances will be credited to this claim. The normal direct job expenses and overhead as contemplated in the bid price will be deducted from actual costs, all of which the electrical contractor considers reasonable considering the circumstances and conditions under which the project was built.

In a sense, what is being described is a total cost approach. It should be obvious that it would be extremely difficult, if not impossible, to have proceeded on any other basis as far as direct job expenses and overhead were concerned. The contractor who uses the total cost approach concentrates on identifying the difference between the reasonable cost of the work as bid and the reasonable cost of actual contract performance.

Table 5.5 illustrates the way the computation of rates for allocation of indirect expenses is accomplished, and how the summary of total cost is determined.

Items of *direct job expense* are costs other than the costs of material, labor, and subcontracts which are incurred by the electrical contrac-

TABLE 5.5 Allocation of Indirect Expenses and Summary of Total Cost

Determination of indirect expenses:		
Total direct costs for the home office for year*		$____
Direct costs on the job for the year		$____
Direct costs on the job as a percentage of total direct costs		%
Total indirect expenses for home office for year		$____
Indirect expenses allocable to job: XX% times		$____
Summary of job costs:		
Total direct costs on job for year		$____
Indirect costs on job for year		$____
Total costs for the job		$____
Determination of overhead expenses:		
Indirect expenses by percent allocation (from above)		$____
Normal contemplated overhead		
Original allowance in estimate	$____	
Plus approved modifications	$____	
Overhead allowed		$____
Net extended and excessive overhead		$____
Determination of direct job expense:		
Total audited expense for duration of job		$____
Normal contemplated job expense	$____	
Plus approved modifications	$____	
Job expense allowed		$____
Net extended direct job expense		$____
Gross total of overhead and direct job expense		$____
Profit at 10%		$____
Total overhead and direct job expense		$____

**Note:* "For the year" is actually the duration of the job. The period of costs for the home office and the job are identical periods of time.

tor, and which result directly from a particular contract. Direct job expenses include the following:

- Payroll taxes
- Insurance
- Fringe benefits
- Equipment depreciation costs
- Subsistence for out-of-town employees
- Performance bond
- Equipment rental
- Field supplies
- Expendable tools
- Job office supplies and expenses
- Job telephone and telegraph

Transportation expenses for personnel and materials

On-the-job truck expenses

Travel of personnel to out-of-town job

Storage and job offices

Temporary power costs

Reimbursement of Borrowed Capital

Almost every electrical contractor has several jobs in progress at the same time. The economics of the business, coupled with the relative size of the electrical job (on average, 15 percent of the total general contract), necessitate this in order to accomplish a sufficient volume of work to provide a profit for the electrical contractor.

When planning the volume of business required to satisfy the budget, and to conform with the available operating capital of the business, it is necessary to include some small and some large jobs in order to arrive at a suitable "mix." However, if in the course of the year's work some of the jobs involve a disproportionately large number of Change Orders of substantial magnitude, the contractor may have to supplement his working capital with bank loans in order to meet the financial requirements of the work. It is possible for a contract to double in size, with the customer demanding that it be completed within the same time frame specified in the original contract. Needless to say, this will stretch the financial capabilities of most firms when the job is a particularly large one.

Clearly, it takes considerably more working capital to finance a job costing several million dollars than one costing half as much in the same period of time. Presumably, any financing cost required to perform the original contract will be included in the original bid, and the question at issue will concern only the additional financing charges. While the amount of additional working capital needed may have been increased because of the failure of government to provide proper payments as the work progressed, this cannot be considered to negate the basic premise: that as significant additional work is added within a given performance period there must be a corresponding increase in working capital. As additional working capital is required, additional interest must be paid, as shown in Table 5.6.

It stands to reason that interest expense for working capital borrowed for the performance of ordinary construction work is not reimbursable when there are no limitations on the time in which the work must be performed. However, when there is no extension of time allowed to perform an amount of work in excess of that specified in the original contract, then interest on additional borrowed working capital is a compensable expense.

TABLE 5.6 Determination of the Cost of Borrowed Capital

Final summary of direct costs	$ ______
Extended overhead addition	$ ______
Prime cost	$ ______
Profit @ 4.8%	$ ______
Total contract price	$ ______
Less: original contract price	$ ______
Amount of contract extension	$ ______
Working capital required @ 1/10 contract extension	$ ______
Interest @ 10%	$ ______

Joint Occupancy with Construction Forces Other Than Your Own

The term *joint occupancy* is used to indicate that the work was performed under limited-space conditions with construction forces other than your own, creating a working condition which differed from those represented by the bid documents. Such restricted-workspace conditions have negative consequences, as follows:

1. The prolonged presence of an excessive number of workers who occupy the same limited space creates a condition for reduced productivity.
2. Normal working conditions are based on having unrestricted floor areas with minimal obstructions to move or work around, including people, unless the specifications indicate otherwise.

When preparing an estimate for bidding on a job, working conditions must be evaluated very carefully because they have considerable bearing on the efficiency which can be expected on the job. *Efficiency* is defined as the capacity to perform a given task in a specified time, at a reasonable cost, and with a minimum of waste. The factor that determines the amount of the joint occupancy claim is the loss of efficiency of the work force due to their inability to work effectively in a confined space with the limitation forced upon them by having conflicts beyond those represented by the specifications. Therefore, it is essential to determine the degree of loss of efficiency of the work force, and in this case it is entirely a question of the rate at which construction is completed compared with the rate that was considered ideal under the conditions which were represented in the specifications.

In order to determine the loss of efficiency, it is first necessary to prepare a construction progress chart such as is shown in Fig. 5.3, where the solid line represents the adjusted schedule, the dashed line shows the actual progress, and the dotted line shows the scheduled progress as adjusted to allow for time losses resulting from joint occupancy. From this chart is determined what the percent completion was

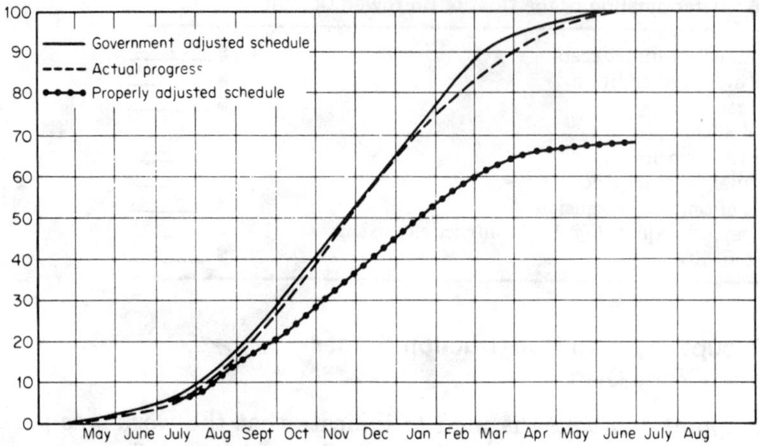

Figure 5.3 Construction progress chart.

by the distances between the lines on the chart. For each month the efficiency loss is determined by the following ratio:

$$\text{Efficiency loss ratio} = \frac{\text{actual percent complete (adjusted)}}{\text{scheduled percent complete in original plan}}$$

The man-hours lost are determined by multiplying the efficiency loss ratio by the actual man-hours worked during each month. The man-hours lost each month are totaled and the cost estimate shown in Table 5.7 is developed.

TABLE 5.7 Cost Estimate and Price Proposal for Overloading

Labor cost:	
Premium pay for overload time	$_____
Overtime efficiency loss: X hours @ $XX.XX per hour	$_____
Overload efficiency loss: X hours @ $XX.XX per hour	$_____
Total added labor cost	$_____
Less: Acceleration previously charged	$_____
Net labor charged	$_____
Other direct costs:	
Top supervisor @ 10% of added labor cost	$_____
Small tools and supplies @ 7% of added labor cost	$_____
Overhead:	
Job overhead @ 15%	$_____
Home office overhead @ 4.5%	$_____
Total cost	$_____
Profit @ 10%	$_____
Total	$_____

Time Extension and Liquidated Damages

It is essential to determine the number of days extension of time for the contract because there is usually a clause for liquidated damages to which the electrical contractor may be subjected. The extension of time is based on the same provocation as the efficiency loss percentage applied to manpower overloading.

On the basis of an average of 22 working days per month, and using the same efficiency loss percentages as were used for the joint occupancy determination, Table 5.8 shows the way in which the days lost are determined.

Summary

A summary of costs resulting from claims is shown below:

Acceleration and decreased productivity*	$________
Additional engineering costs	$________
Extended overhead and direct job costs	$________
Financing the job with borrowed capital	$________
Joint occupancy due to manpower overloading	$________
Time extension and liquidated damages	$________
Total due on claims	$________

*Time Extension: 69 working days.

It is not sufficient merely to recognize the cause of a loss which would result in claims and then to present the claims for reimbursement; the cause must be properly evaluated and completed supported by documentation. Even after this is done, negotiation will be re-

TABLE 5.8 Determination of Days Lost

Month	Days per month	Efficiency loss percentage	Days lost
July	22	23	5
Aug.	22	31	7
Sept.	22	34	7
Oct.	22	26	6
Nov.	22	31	7
Dec.	22	40	9
Jan.	22	36	8
Feb.	22	31	7
Mar.	22	26	6
Apr.	22	21	5
May	22	11	2
Jun.	22		
Total days lost			69

quired and the amount paid will differ from that requested. The process of negotiation involves a certain amount of deception and misrepresentation; this should be recognized as a necessary ingredient in the claims procedures, and should be included as an extra item of cost in preparation of the cost proposal.

Chapter

6

Principles of Effective Operations

Success in the electrical contracting business is practically ensured when operations are conducted based on a complete business system. A *system* may be defined as "a regularly interacting or interdependent group of items forming a unified whole." *System* is also defined as "an organized or established procedure." Both of these definitions are relevant to our definition of a complete business system: a *complete business system* may be defined as a series of procedures which are interrelated in such a way that each work process is dependent on the results of the other processes.

Examples of interrelated procedures are the following:

- Estimating and preparing the Bill of Material and purchasing
- Preparing the Bill of Material and job management scheduling
- Job management scheduling and purchasing
- Purchasing and accounting invoice payments
- Job scheduling and determining manpower requirements
- Determining manpower requirements and determining payroll requirements

All the above events are interrelated and depend on the way in which the estimate is prepared and the associated paperwork which relates to the estimate. Like the instruments of a symphony orchestra, all the components must be interphased for perfect continuity and harmony in order to be effective.

Each of the processes in electrical contracting occurs in an orderly, sequential manner, and each is very much dependent on those which preceded it. This makes the establishment of a complete business system even more valuable as a management tool for this particular business. Note that in such a system each process is performed by a specialist who is independent of the other functions in the chain of operations, but whose work product is used to provide the basis of their work ethic.

Operations in the electrical contracting business consist primarily in estimating and job management.

Estimating is the process of (1) forecasting future costs of labor, material, and direct job expense by envisioning from plans and specifications the conditions under which the installation will be made, and then (2) counting, measuring, and listing the items of material required to make the installation; assigning a standard labor unit to each item of material; and factoring the labor units as required so they will conform to the conditions of the job. When the figures obtained are summarized with costs of overhead and profit, estimating will result in a competitive bid for a job which can be installed at a profit.

Job management is the process of providing maximum field support in order to encourage and obtain the best possible productivity; "support" manifests itself in good supervision, instruction, and guidance such that work is installed in accordance with the terms of the contract and within the framework of the estimate.

The mastery of these two functions is indispensable to the success of an electrical contracting business, and virtually all the functions in the business are dependent for proper execution on the form and structure of the estimate; therefore, great care must be taken to understand and execute the requirements of an estimating procedure that will produce the desired results.

Experience indicates that the use of standardized estimating is essential for speed, accuracy, and consistency. Standardized estimating has the following benefits:

- The estimator has increased confidence in his work.
- The speed of estimating is increased.
- Omissions are minimized.
- Uniformity results in better communications between estimators.
- Job files are more uniform and more intelligible.
- Communications with purchasing and warehousing are established.

The standardized estimate must be flexible because the requirements of jobs vary so widely; however, most variations can be reconciled through the use of proper estimating forms. Proper estimating forms provide flexibility to accommodate labor unit adjustments and factoring without sacrificing the desirable features of standardization.

Standardized estimating is the process of applying the same basic estimating procedure to all jobs of all types using standard labor units which are applied to items of material, and then adjusting the labor units according to the conditions of the particular job and factoring to compensate for differences in conditions of usage and installation. Regardless of the size of the job, every estimate will cover the same divisions of work, and the same forms can be used for all estimates. It is essential that work be taken off the blueprints in a manner that conforms with the way in which the labor units are structured, and the estimating forms are designed to accommodate the takeoff. In addition to providing a format for adjusting the basic labor units, proper forms provide a sequential outline of procedure which acts as a work organizer for the estimator. They should therefore result in fewer oversights and errors, with the ultimate results that the estimator has more confidence in his work.

The ultimate objective of every estimator is to achieve the best possible success ratio in bidding. The actual success ratio of 1:7 is a conclusive indication that a great deal of the estimator's time is nonproductive in terms of producing successful job estimates—which simply emphasizes the fact that the secret of success is to employ an appropriate estimating system which is thoroughly understood and properly applied.

There are 13 principles of operations that we can apply to the sequence of functions to be discussed in this chapter. Each provides a link in the chain of procedures required to form a complete business system for the electrical contractor.

Principle 1: Establish Conditions of Installation

The concept of conditions of installation underlies the entire business system. This significant statement is predicated on the fact that the procedures used in estimating are determined by the conditions of installation and that the estimate provides the basis for organizing nearly all the functions performed in the electrical contracting business.

We have said that estimating is the process of forecasting the cost of installation of electrical work in a structure by envisioning from plans and specifications the conditions under which the installation will be made. Since there are so many different types of structures, building materials, heights of installation, and positions with respect to the ground, it has become apparent that it is necessary to establish a set of "basic" conditions which normally occur in the majority of installations. When the plans indicate conditions different from the normal, it becomes necessary to deviate from the basic conditions by applying adjustments to the labor units of the materials required for the installation. The basic, or normal, conditions are called the *conditions of installation,* and have been established for all classes of material for which a standard labor unit has been assigned. A *standard labor unit* may be defined as "the man-hours required to handle and install an item of material in certain types of construction."

Uniform conditions may be established as follows:

- A journeyman of average ability using normal tools and procedures will make the installation.
- The material will be installed in a localized area of approximately 40,000 sq ft.
- The height of installation will not exceed 14 ft above the floor.
- Access to the work will be normal for new construction, and it will not be necessary to move obstacles on the floor.
- The work will be installed at ground level or similar access.
- The work will be installed by one man as required for one-man jobs, or two men as required.
- The rate of progress of construction will be normal.
- Temperature conditions will be normal and special clothing will not be required.
- The concentration of material in a localized area will be normal.

The use of these conditions of installation determines the estimating procedure, because the use of standard labor units (after the appropriate adjustment has been applied) will apply to all work which meets these conditions regardless of the location of the work in the structure. However, the estimator must recognize from the conditions represented on the blueprints all deviations from standards that will occur under the actual construction conditions. The deviations must be recognized as the estimate is being prepared, and the deviations must be flagged on the takeoff forms in order to apply the proper factor on the pricing form.

Thus, to repeat, the conditions of installation form the basis for the standard estimating procedure, and for the intelligent application of job factors. Let it be understood that when construction conditions are equal to or more favorable than the conditions of installation, the standard labor units will be applied to items of material taken off in the usual manner without factoring, but when conditions are less favorable than those specified it is necessary to adjust the units accordingly.

Adjustments for conditions of installation

The following are conditions of the job that can differ from the conditions of installation, and that therefore require adjustment of the standard labor unit when estimating:

- Substandard labor affecting productivity
- Labor agreements that require two men working together for all operations
- Work installed more than 14 ft from the floor
- Length of wire pulls and of wire pulls in parallel
- Conduit runs in parallel
- Preassembly and prefabrication
- Large concentration of items in a localized area
- "Checkerboarding" of the job by the general contractor, or the rate of progress on the job that is not normal, or the job is rushed
- Work done over obstacles on the floor
- Severe weather conditions, hot or cold
- Material furnished by others than the contractor
- Conditions that require the workmen to wear special clothing

Principle 2: Establish Standard Labor Units

The processes of estimating involve counting and measuring items of material and applying to each a standard labor unit in man-hours. This standard labor unit will represent the time spent in the field to handle and install the items of material in accordance with the conditions of installation and the conditions of usage (see under "Principle 3: Conditions of Usage").

Distribution of time

The value of the labor unit is determined by analyzing the *distribution of time,* that is, the portion of total in-the-field time required by the field men to perform each of the eight time-consuming activities involved in making an installation. The percentage distribution of field time is detailed below:

Operation	Average percentage, all classes of material
Studying plans and specifications	5.0
Ordering material and tools	2.0
Receiving and storing material and tools	4.0
Moving material to the point of usage	6.0
Tooling up	6.0
Measuring and laying out the work	4.0
Installing	63.0
Performing nonproductive tasks	10.0
Total	100.0

It should be emphasized that these percentages represent an average of all the percentages for the many classes of material involved in making an installation. Percentages vary widely for the different classes of material. For example, the tooling factor for installing feeder wire is 28 percent, whereas the average tooling factor shown in the table above is 4 percent; the installation factor for feeder wire is 43 percent, whereas the average for this factor as shown in the table is 63 percent. (The feeder wire example shows the widest deviation from the average of any class of material, but is therefore a good example to illustrate the potential for substantial differences between the classes of material.) Table 6.1 further illustrates the differences in the distribution of time for several of the different classes of material.

An analysis of the factors influencing the eight time-consuming activities provides an understanding of how field time may be influenced by good management on the part of the contractor, and also shows the importance of planning and scheduling on the part of both the job management team and the journeymen on the job. The time-consuming functions are discussed individually in the sections that follow:

Studying the plans and specifications (2 to 10 percent of total time). The amount of time spent studying the plans by the foremen in the field depends largely on how well the electrical work is laid out on the plans, and how well information relative to the details of the work is transmitted by the contractor to them. It is essential for the circuiting to be clearly defined on the electrical drawings, or the foreman will have to spend time doing this. Such information as panel schedule and fixture schedules is essential for speed of interpretation. If complicated control information appears on more than one plan, it must be consolidated for speed and accuracy to a single plan. If the foreman on the job is not furnished with a Bill of Material, it will be necessary for

TABLE 6.1 Distribution of Time for Various Classes of Material (In Percent)

Operation	Class of material								
	Building wire	EMT	GRSC	Safety switches	Branch panelboard	GRSC fittings	Lighting fixtures	Aluminum busway	4" square outlet box
Study plans	2	7	4	4	8	3	5	3	8
Order material	2	2	2	2	2	3	2	2	2
Receive and store	5	5	4	3	4	5	5	3	3
Move to usage	5	4	4	3	4	5	5	3	5
Tool up	28	8	10	5	3	15	5	25	5
Measure and locate	3	6	4	5	3	5	5	5	10
Install	45	58	62	68	66	54	63	49	57
Nonproductive	10	10	10	10	10	10	10	10	10
Total	100	100	100	100	100	100	100	100	100

him to take off a list of the material items required and order them from the contractor's office or a designated supplier.

It is important in the interest of time to have an adequate number of copies of plans on the job, and to maintain them in a proper manner. Plans used in the job field office should be maintained on plan holders and stored on racks. Plans used in the work areas should be mounted on plywood or masonite, and the edges should be bound with mastic tape.

A minimum of time should be spent by journeymen studying the plans—this must be done only by the foreman. When the journeymen are given a work assignment, they should be given the required plan for the work and it should be reviewed by the foreman with them.

Preparing blueprints properly and employing suitable construction techniques can substantially reduce the field time spent on this activity.

Ordering material and tools (1 to 5 percent of total time). It is essential for the foreman on the job to be furnished with a Bill of Material by the contractor. The Bill of Material must be subdivided into categories of work according to the stage of construction in which it is used, thus eliminating the need to stock the job at all times with all the items required for the total job.

In fact, it should only be necessary for the foreman to request material from time to time, and then only from the classes of material required for the particular stage of construction then ongoing.

If a Bill of Material is not prepared in advance for field usage, it may be necessary for the foreman to lay out the circuiting and take off the quantity and type of material; this is a time-consuming operation and one which has already been done in the process of estimating. Thus, an essential part of preparing the estimate is to provide for subdividing material needs in accordance with the stages of construction.

Receiving and storing material and tools (1 to 5 percent of total time). It is difficult to precisely coordinate the need for material with its arrival on the job. In fact, since it is good management practice to have the material on hand on the job before it is actually required, it is generally necessary to store material at the job site in advance of its use.

The time spent performing this operation can be greatly minimized by providing proper storage facilities for the material in close proximity to the work. Truck trailers or job warehouses may be properly placed and organized so that material items can be maintained in separate containers and easily located.

A considerable amount of time may be saved if the job facilities are clean and orderly. For long-duration jobs this is an important factor. In addition to the time that can be wasted trying to find an item of material, there is often an improper utilization of capital as a result of carrying excessive amounts of material which are never used. But while it is desirable to keep job stocks as small as possible, it is also desirable to make as few deliveries to the job as possible. Material should be stored as close to the point of usage as possible, and lockboxes for the journeymen's tools should be correspondingly close.

The time allotted for this category also includes removal of material and storage facilities from the site at the end of the job.

Moving material to the point of usage (1 to 5 percent of total time). The efficient utilization of time in executing this function depends on the employment of proper conveyance equipment, the proper placement of points of storage, and adequate storage facilities. The required items of material must be readily accessible and easily transferable to a container which can be transported by the journeyman to the point of usage. The following equipment makes it possible to accomplish this:

- Tote boxes for material and tools
- Four-wheel hand carts with large casters and shelves
- Two-wheel trunk carriers with large wheels
- Removable storage boxes that are portable
- Multilevel racks with multiple storage boxes
- A four-wheel heavy-duty truck
- Tool lockboxes
- Material lockboxes

A Bill of Material which is listed according to areas, floors, systems, or specialized functions such as feeder runs makes it possible for a journeyman to easily discern the items of material required for a certain work assignment.

Packaging and bundling are important considerations in matters pertaining to handling and storing material on the job. For the most part, the manufacturer is responsible for this, but many times the form of bundling can be specified at the time of purchase to accommodate the requirements of the job.

Tooling up (1 to 35 percent of total time). Many of the principal tools used on the job, such as electric drills, electric hammers, and benders should be housed in their own boxes, and all tools and equipment should be identified by stenciling and painting so that they may be quickly identified when required. Loss of time can be avoided by servicing all tools periodically to ensure good operating condition when the tools are needed. Proper tooling has such a profound effect on the man-hours required to install electrical work and perform all the associated processes that a great deal of attention must be given the modernization and upgrading of equipment used on the job, including scaffolding, hoists, trenching equipment, and wire-pulling equipment.

One of the aspects of tooling up is providing temporary power when it is required. Most jobs are provided with a network of temporary power circuits, but if these are not available it may be necessary to use a portable generator.

Setting up a tool often involves extensive bracing or excavation. This should be avoided, if possible, by selecting a tool which does not require this time-consuming operation.

The utilization of certain tools for certain operations is part of the established conditions upon which the standard labor units are based. It is essential to use the proper standardized tool in accordance with the estimating standards, or else the labor hours estimated will be at variance with the actual time spent.

Measuring and laying out the work (2 to 25 percent of total time). In order to ensure accuracy and effective use of time in the field, it is essential to provide the job with a set of properly prepared electrical construction drawings which include all the objects which require electrical connections as shown on the architectural and mechanical plans.

These drawings should include the following:

- Complete circuiting of branch circuit work
- Feeder layouts and pulling schedule
- Fixture schedule with hangers and lamps
- One-line diagram
- Schedule of finishes of switch and receptacle plates
- Panelboard load schedules
- Field installation details and sketches
- All partitions located by dimensions
- All architectural materials shown by symbols

In addition, the journeymen must have all the manufacturers' installation instructions; these generally accompany the material or equipment when it is delivered to the job.

Installing (35 to 75 percent of total time). The process of installing involves putting the item in place and completing all supplementary operations such as those required for wire terminals, trims, covers, and directories. The labor units do not always include these supplementary operations, and therefore each item of supplementary material must be defined. The expendable materials are included in the material cost as a function of labor cost in the amount of 0.2 percent.

The following factors affect the value of the labor unit.

- Weight and bulk
- Area and volume
- Amount of assembly required before installation
- Packaging, including boxing and bundling
- Type and size of tools required
- Access to work, involving type of construction
- Position of installation relative to the floor
- Number of men required to make the installation
- The rate of work accomplishment by the journeyman
- Supplementary operations required, such as bending or offsetting

Not all these factors are involved in the determination of each labor unit, but all the labor units are affected in some way by some of the factors.

Substantial savings in man-hours may be realized by selecting materials which require minimum assembly or fabrication before installation as well as by selecting materials of reduced size and weight. Prefabrication and production line assembly methods are effective in reducing man-hours because of the concentration of work and improved tooling methods.

Performing nonproductive tasks (10 percent of total time). All the direct labor time spent on the job which does not relate to handling or installing material is included in this category, as follows:

- Lost time
- Toilet time
- Smoking time
- Coffee-break time
- Conversation time
- Time spent removing paper and wood cartons from premises
- Time spend in sweeping and cleanup of the premises
- Lost motion

These factors occur on every job every day, but there are also other factors which may result in considerable lost time. This lost time is inexcusable because these factors are controllable by the supervision of the job. They are:

- Late starts and early stops in observing work hours
- Unnecessary interruptions
- Extended lunch hours and coffee breaks
- Wasted time (due largely to poor work assignments)
- Excessive conversation time
- Unnecessary duplication of steps
- Failure to plan ahead
- Working below capacity
- Having too many journeymen on the job

None of these items should be included in the standard labor unit because they are controllable and should not be permitted.

Application of adjustment factors to the standard labor unit

The importance of fully understanding the distribution of time for the eight elements of time-consuming labor in the field cannot be overemphasized, because these elements provide the basis for the formation of all labor units, and enable the estimator to resolve the deviations from standards which occur in every estimate.

While standard labor units represent the bulk of items involved in most wiring jobs, almost every job introduces items for which the estimator must create labor units, and a thorough understanding of the elements of time-consuming field labor is essential for doing this.

In addition to furnishing the information for the formation of the basic labor units, the eight elements provide the fundamental information which is required to factor for the following conditions:

- Length of wire pulls
- Parallel pulls
- Parallel runs of conduit
- Concentration of items of material and equipment installed in a single location
- Two-man versus one-man operations

No supervision time is included in the basic labor units. On jobs requiring foremen and general foremen, the time for supervision is established by using the distribution-of-time data to determine the amount of productive time for each supervisor. Productive time is included in the standard labor units, and for supervision it is a function of the number of men under supervision. The number of men under supervision is determined by preparing a Manpower Chart, and recognizing the labor agreement.

The material used by the electrical contractor on a job is almost entirely a function of the type of construction, and labor units differ for the different types of construction. It is therefore essential that the estimator recognize the type of construction when evaluating the labor unit which is assigned to the material. The following examples illustrate how the type of construction influences the value of the labor unit for the same item of material:

Item of material	Type of construction	Labor unit
Box, 4" octagonal	Exposed	0.16MH ea
	Furred	0.19MH
	Masonry flush	0.45MH
Box, 4" square	Metal frame	0.22MH ea
	Masonry flush	0.45MH
	Exposed	0.16MH
EMT conduit, ½"	Deck, turned up	0.031MH/ft
	Exposed	0.027MH/ft
	Notched joist	0.030MH/ft

When branch circuit conduit is installed in various types of construction, the cost of installation is affected by the number of bends and offsets per 100 ft which are required to fit the work into the different types of construction, as follows:

Types of construction	Number of bends and offsets/100 ft
Deck	Two bends and two offsets
Exposed	Four bends and four offsets
Furred ceiling	Two bends and two offsets
Metal frame	Three bends and three offsets
Wood frame	Four bends and four offsets
Conduit on hangers	Two bends
Flush in masonry	Three bends and three offsets
Trench	Two bends
Masonry	Three bends and three offsets
Concrete walls	Three bends and three offsets

Each bend and offset involves 0.05 man-hours of labor. The standard labor unit per conduit is based on free air construction. When these units are applied to the various types of construction, the mul-

TABLE 6.2 Application of a Multiplier Factor to the Standard Labor Unit

Type of construction	Multiplier*
Deck	1.20
Exposed	1.40
Furred ceiling	1.30
Metal frame	1.30
Wood frame	1.50
Conduit on hangers	1.10
Flush-in masonry	1.45
Trench	1.20

*Including access factor (see Table 6.3).

tiplier for branch circuit conduit labor units is established as shown in Table 6.2.

In addition to the increased man-hours caused by the need for bends and offsets, there is an additional factor called the *access factor* which applies to branch circuit work (see Table 6.3).

The lists in Tables 6.2 and 6.3 are representative, but not complete. Many different types of construction are involved in the building of facilities of different kinds, and each affects the basic labor unit, and therefore the total man-hours of labor, differently.

Definitions of and classifications for the different types of construction are shown in Fig. 6.1.

The Labor Unit Tables shown in the Appendix are designed to provide information about the following:

- Standard labor units with values in man-hours
- Adjusting labor units by applying a factor according to conditions which differ from the conditions of installation
- Conditions of installation
- Conditions of usage
- Distribution of time

TABLE 6.3 Access Factors for Branch Circuit Work

Type of construction	Multiplier
Metal frame	1.05
Flush-in masonry	1.15
Furred ceiling	1.10
Wood frame	1.10
Trench	1.10
Exposed	1.10

FREE AIR

FOR ASSEMBLIES WHICH DO NOT REQUIRE FASTENINGS.

SLAB

FOR CONDUIT ASSEMBLIES CONCEALED IN CONCRETE FLOOR OR CEILING AND STUBBED UP OR DOWN.

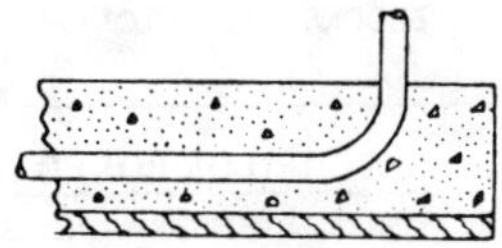

EXPOSED ON CONCRETE

FOR ASSEMBLIES INSTALLED ON THE SURFACE OF SOLID CONCRETE, BRICK, STONE, ETC.

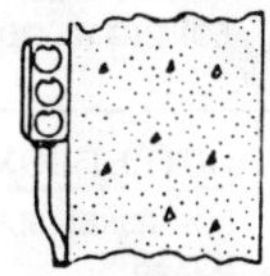

DECK FLUSH

FOR ASSEMBLIES CONCEALED IN CONCRETE SLABS, EITHER CEILING OR FLOOR.

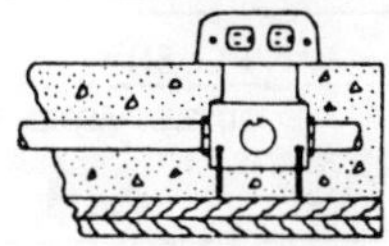

DRAWN IN CONDUIT

FOR WIRE ASSEMBLIES DRAWN IN CONDUIT.

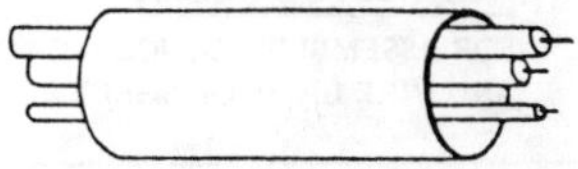

EXPOSED ON WOOD

FOR ASSEMBLIES INSTALLED ON THE SURFACE OF WOOD.

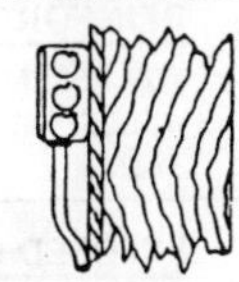

FURRED FLUSH

FOR ASSEMBLIES INSTALLED ABOVE SUSPENDED CEILINGS.

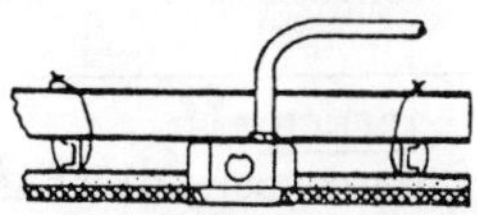

EXPOSED ON HOLLOW MASONRY

FOR ASSEMBLIES INSTALLED ON THE SURFACE OF HOLLOW MASONRY.

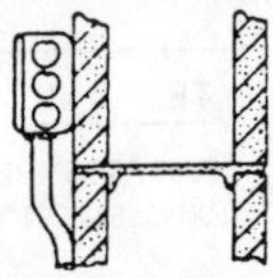

JOIST CONSTRUCTION

FOR ASSEMBLIES EXPOSED ON STEEL JOISTS.

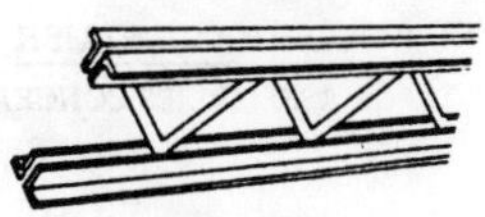

LAID IN WIREWAY, CHANNELS, ETC.

FOR WIRE ASSEMBLIES LAID IN GUTTER, WIREWAY, OR IN FIXTURES MOUNTED IN CONTINUOUS ROWS.

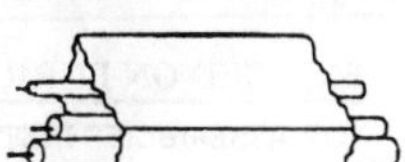

Figure 6.1 Types of construction. (*Estimatic Corp.*)

MASONRY FLUSH

FOR ASSEMBLIES CONCEALED IN BRICK, CINDER BLOCKS , TILE, ETC.

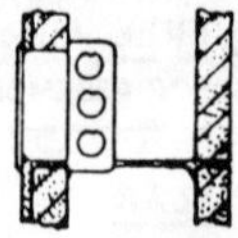

NOTCHED OR BORED JOIST FLUSH

FOR CONDUIT OR WIRE ASSEMBLIES CONCEALED IN WOOD FRAME CONSTRUCTION BY NOTCHING OR BORING WOOD MEMBERS.

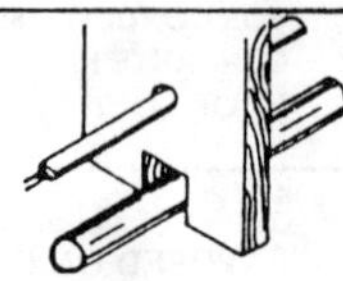

OUTLET BOX MOUNTING

FOR ASSEMBLIES INSTALLED DIRECTLY ON OUTLET BOXES.

PAN FLUSH

FOR OUTLET ASSEMBLIES CONCEALED IN CONCRETE BEAMS.

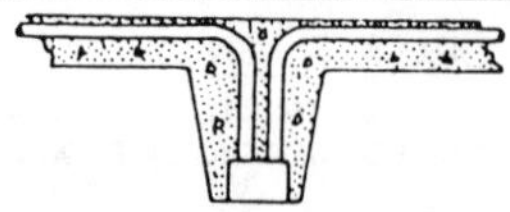

METAL STUD FLUSH

FOR ASSEMBLIES CONCEALED IN WALLS OF PLASTER AND WIRE LATH, OR SHEET ROCK, ON METAL STUDS.

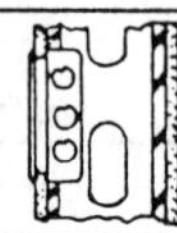

REINFORCED CONCRETE FLUSH

FOR ASSEMBLIES CONCEALED IN REINFORCED CONCRETE WALLS.

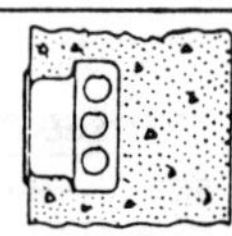

EXPOSED ON STEEL

FOR ASSEMBLIES INSTALLED ON THE SURFACE OF METAL.

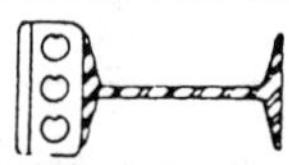

TRENCH

FOR CONDUIT OR WIRE ASSEMBLIES CONCEALED IN THE EARTH. (DOES NOT INCLUDE TRENCHING OR BACKFILLING).

TEE–BAR

FOR ASSEMBLIES INSTALLED EXPOSED ON TEE-BAR.

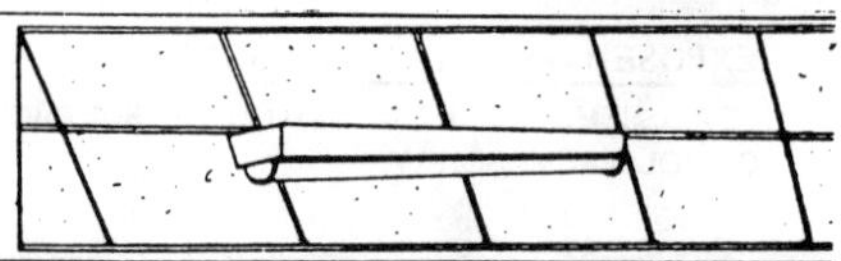

WOOD FRAME FLUSH

FOR ASSEMBLIES CONCEALED IN WOOD FRAME WALLS, CEILINGS, OR FLOORS.

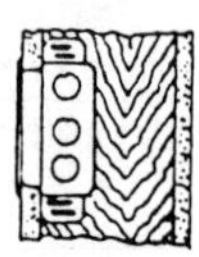

EXPOSED ON FURRED

FOR ASSEMBLIES INSTALLED ON THE SURFACE OF SUSPENDED CEILINGS.

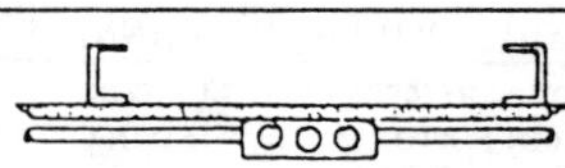

Figure 6.1 *(Continued)*

Knowledge of each of these is essential to the proper application of the labor unit, and it is mandatory for the estimator to have a thorough understanding of the standard conditions in order to apply the labor units to all types of construction with required accuracy. The principle is to apply the standard labor unit and then make the necessary adjustments by using a multiplier to increase or decrease the unit depending on the way in which the conditions of the job at hand differ from the standard conditions.

Such adjustments to the labor units are often numerous and voluminous, and it is difficult to handle them in an orderly manner unless the estimator uses forms and estimating procedures which are designed to facilitate these adjustments. The design of the forms must be such that the adjustments can be made easily and accurately. This is best accomplished with complete coordination of forms, estimating procedure, and compilation of labor units. In addition, the entire set of estimating forms must be interrelated in terms of application, factoring, and adjustment of labor units.

Principle 3: Establish Conditions of Usage

A *standard labor unit* is defined as the man-hours required to handle and install an item of material in accordance with the conditions of usage and the type of construction in which it is placed. The term *conditions of usage* relates to the application of labor units in accordance with the conditions under which the material is installed in the job.

We have emphasized that an important part of the concept underlying all estimating procedures is the use of standard labor units for all estimating, followed by adjustment of those standard units in accordance with the construction conditions with which the estimator is working. The conditions surrounding the standard labor unit are clearly defined in the Labor Unit Tables in the Appendix for each item of material, and the description of the manner in which the units are adjusted is given by the conditions of usage.

Some of the typical classes of material, with their conditions of usage, are the following:

Feeder wire	Standard labor units for feeder wire are based on pulls of 100 ft for each of the numbers of conductors from 1 to 9. Pulls of greater or lesser lengths must have their units adjusted according to the length of run.
Feeder conduit labor	Standard labor unit includes one bend per 200 ft. No fastenings, ells, or fittings are included.
Conduit terminals	The labor unit is applied to the bushing, which represents three-fourths of a cut and thread on

	rigid conduit. No conduit terminals are included in the labor unit for boxes or panels.
EMT conduit terminals	Each connector represents a terminal, which includes cutting, measuring, and deburring. The value of the terminal varies with the type of construction because some terminals involve bending or offset.

Each class of labor involves adjustments, and the adjustments are described in the Labor Unit Tables. The calculations for determining most of the adjustments are based on distribution-of-time tables. A typical calculation for feeder cable would be the following:

Example A standard unit run of 100 ft of 500-MCM-type THHN 4 conductor would have a unit value of 0.0525 man-hours per ft. Therefore the total for 100 ft would be 5.25 man-hours, but the 100 ft of 4 conductor represents a total of 400 ft of cable at 0.0525, equaling 21 man-hours total. The total for a 200-ft pull without adjustment would be 10.50 man-hours, but with adjustment is 7.72 man-hours, which is 26.5 percent less. However, that represents 800 ft @ 0.038, equaling 30.88 man-hours, instead of 42 man-hours, which would be the amount without factoring.

The distribution-of-time table on which this calculation is based is shown below:

Description of activity	Normal, percent	Percent on extra 100 ft
Study plans	2	0
Order material	2	0
Receive and store	5	1
Move to point of usage	5	1
Tool up	28	0
Measure	3	1
Install	45	45
Nonproductive	10	5
Total	100	53

Therefore there is a reduction of time on the additional 100 ft of 47 percent of the man-hours required to handle and install the wire, which will result in total man-hours as follows:

Man-hours required to handle and install the basic 100 ft	5.25
Man-hours required to handle and install the additional 100 ft	2.47
Total man-hours for the 200-ft pull	7.72

As noted in the example, these figures indicate a reduction in total time of 26.5 percent with factoring, or an adjustment factor (multiplier) of 0.735.

If there were two runs in parallel, the additional adjustment factor would be 0.90, which is calculated in a similar manner to the extra 100 ft.

Assuming there are two runs in parallel of 200 ft each, the comparison of the estimate with factoring to that without factoring is:

Total man-hours unadjusted = 2 × 10.50	21.00
Total man-hours with adjustment factor applied	13.90
Difference between unadjusted and adjusted man-hours	7.10

The percent reduction in this case is 33.81.

This example shows the results of estimating just one feeder; it clearly shows that it would be impossible to be competitive on a sizable job without recognizing the true values.

It must be understood that the conditions of usage pertain to the application of labor units to the individual item of material in accordance with the conditions of installation, and the type of construction in which the installation is being made.

Adjustment factors should be applied to standard labor units under the following conditions:

1. *Quantity of items for normal concentration of work:* Standard labor units may be reduced considerably when there is a high concentration of work in a localized area. Table 6.4 shows the percentage by which the standard labor units may be reduced when concentrations of given number of items occur in a localized area.

2. *Manpower work assignments for specific classes of material:* Table 6.5 has been prepared to show the number of men assumed when preparing the standard labor unit for various classes of material. This table shows the one-man and two-man installation assignments for the different classes of material.

When two men are used to make the installation instead of the one man upon which the labor unit is based, the actual cost is increased 65 percent.

3. *Height factor for work installed more than 14 ft above floor:* For installations more than 14 ft above floor level, it is advisable to use two men to make the installation regardless of the type of equipment used.

TABLE 6.4 Concentration Quantities*

Item of material	Normal quantity	Type of concentration	Percent reduction
Breaker, molded case	4	Concentration of 12	14
	4	Concentration of 24	17
Busway, feeder	100 ft	Over 100 ft	29
	100 ft	Over 600 ft	36
Busway, plug-in	100 ft	Over 100 ft	29
	100 ft	Over 600 ft	36
Cabinet, telephone	4	Concentration of 12	14
	4	Concentration of 24	17
Clamp, conduit	12	Concentration of 24	23
	12	Concentration of 36	29
Cutting, holes	6	Concentration of 18	14
	6	Concentration of 36	17
Duct, surface	100 ft	Concentration of 300 ft	22
	100 ft	Concentration of 600 ft	27
Duct, under floor	300 ft	Concentration of 900 ft	22
	300 ft	Concentration of 1800 ft	27
Elbow, busway	12	Concentration of 36	23
	12	Concentration of 72	29
Elbow, conduit	18	Concentration of 36	21
	18	Concentration of 72	31
Electric heat, baseboard	2	Concentration of 6	11
	2	Concentration of 12	14
Electric heat, sill	2	Concentration of 6	11
	2	Concentration of 12	14
Fitting, conduit	12	Concentration of 36	23
	12	Concentration of 72	29
Fixture, fluorescent	50	Concentration of 72	6
Base: run of 4	50	Concentration of 144	10
Fixture, fluorescent tandem	50	Concentration of 72	6
Base: run of 4	50	Concentration of 144	10
Fixture, incandescent	12	Concentration of 36	16
	12	Concentration of 72	19
	12	Concentration of 108	21
Hanger, fixture	12	Concentration of 36	16
	12	Concentration of 72	19
Housing, meter	4	Concentration of 12	14
	4	Concentration of 24	17
Outlet, extension	24	Concentration of 72	14
	24	Concentration of 144	17
Outlet, fixture	24	Concentration of 72	14
	24	Concentration of 144	17
Outlet, junction	24	Concentration of 72	14
	24	Concentration of 144	17
Outlet, telephone	24	Concentration of 72	14
	24	Concentration of 144	17
Panelboard	4	Concentration of 12	14
	4	Concentration of 24	17
Receptacle, duplex	24	Concentration of 72	14
	24	Concentration of 144	17
Receptacle, single	24	Concentration of 72	14
	24	Concentration of 144	17

TABLE 6.4 Concentration Quantities (*Continued*)*

Item of material	Normal quantity	Type of concentration	Percent reduction
Receptacle, twist type	24	Concentration of 72	14
	24	Concentration of 144	17
Starter, hand	4	Concentration of 12	13
	4	Concentration of 24	17
Starter, magnetic	4	Concentration of 12	13
	4	Concentration of 24	17
Switch, control	6	Concentration of 18	13
	6	Concentration of 36	17
Switch, fusible safety	4	Concentration of 12	13
	4	Concentration of 24	17
Switch, low voltage	6	Concentration of 18	13
	6	Concentration of 36	17
Switch, motor disconnect	4	Concentration of 12	13
	4	Concentration of 24	17
Switch, nonfusible safety	4	Concentration of 12	13
	4	Concentration of 24	17
Switch, tumbler	24	Concentration of 72	14
	24	Concentration of 144	17
Terminal, conduit	100	Concentration of 300	22
	100	Concentration of 600	27
Terminal, wire	100	Concentration of 300	13
	100	Concentration of 600	16
Wireway, screw cover	12 ft	Concentration of 36 ft	22
	12 ft	Concentration of 72 ft	27
	12 ft	Concentration of 144 ft	29

Figure 6.2 shows the effect of ceiling height when different types of support are used. From Fig. 6.2 the following summary may be drawn:

Ceiling height, ft	Adjustment factor
Using Ladder as Support:	
18	1.20
20	1.30
22	1.70
Using Scaffold:	
15	1.20
20	1.50
30	1.70
40	1.90
50	2.60
Using Scissor Lift:	
18	1.10
21	1.40
35	1.50
40	1.90

TABLE 6.5 Manpower Work Assignments

One-Man Work Assignments Include:

- All EMT work ½ in through 3 in, all IMC work ½ in through 2½ in, all GRS work ½ in through 1½ in (except for work on high ceilings or walls), and all polyethylene pipe work up to 3 in
- Operation of all power tools, such as threading machines, hacksaws, and hammers (within the provisions of item 1)
- Work on 4-ft and smaller fluorescent fixtures with two, three, and four lamps
- Work on safety switches of sizes up to and including 200 amp 4SN
- Work on lighting panels up to 32 circuits
- Work on telephone panels with single doors
- Work on all devices for branch circuit outlets
- Work on all incandescent fixtures except chandeliers and multilamp area lights
- Work on motors and controls up to 30 hp
- Work on meter equipment
- All underfloor duct work
- All conduit hanger work
- All work on 16- and 18-ft ladders

Two-Man Work Assignments Include:

- All 4-in EMT work, 3- and 4-in IMC work, and all work on polyethylene conduit 4 in and larger
- All GRS work 2 in and larger
- High-ceiling work using scaffolds, scissor lifts, and 20-ft ladders
- Work on 4-ft fluorescent fixtures with six and eight lamps and work on all 8-ft fluorescent fixtures
- Work involving pulling branch circuits and feeder wire
- Work on safety switches 400 amp and larger
- Work on panelboards 34 circuits and larger
- Work on double-door telephone cabinets
- Work on chandeliers and multilamp area lights
- Work on motors and controls 40 hp and larger

4. *Length of feeder wire pulls:* Standard labor units are established for feeder wire pulls of 100-ft length for number of conductors from 2 to 9. Units are available for sizes and types of insulation in both copper and aluminum.

Pulls of less than 100 ft are factored up. The units for a 50-ft pull are factored by 1.50.

The following pulls of different lengths are factored by numbers which are rounded off, but actually the computer may be used to factor the length of the pull by the foot, which is not practical for estimators using manual methods.

Length of pull, ft	Multiplier
150	0.80
200	0.70
300	0.60

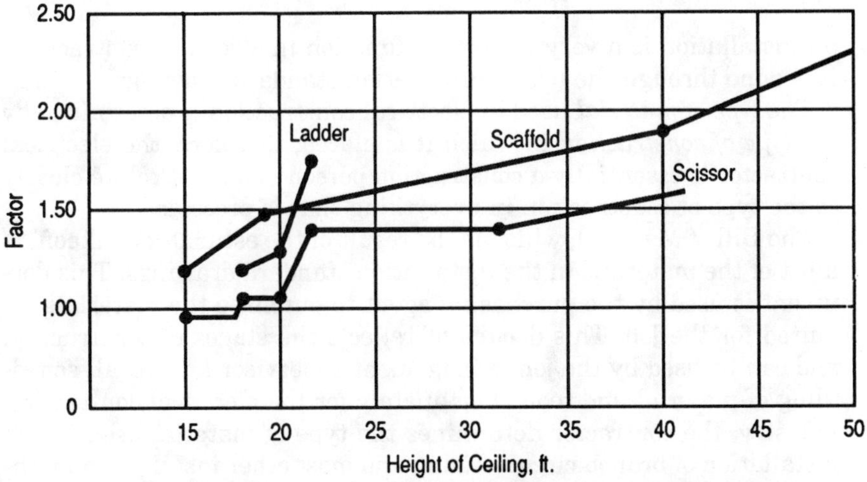

Figure 6.2 Chart showing adjustment factors for installation of work above 14 ft.

The adjustment factor for parallel pulls is predicated on being able to use the same setup and pulling mechanics for each pull.

	Multiplier
Second pull	0.90
Third pull	0.85
Fourth pull	0.80
Fifth pull	0.78

The multipliers for parallel pulls are in addition to the length factoring where applicable.

5. *Conduit runs in parallel:* The reduction in the size of the labor unit which results when conduits are run in parallel is as follows:

Parallel runs of:	Percent reduction
2	17
3	22
4	24
5	26
6	27

There is no reduction for length of run.

Principle 4: Establishing Installation Standards

The coordination of field work with the estimating procedures that involve the determination of the types of materials to be used in making an installation is a very important function in electrical contracting; it is done through the use of installation standard drawings.

The *type of material* used in electrical construction is determined by the *type of construction* in which it is placed; therefore, the electrical contractor is essentially a construction person and must relate closely to the type of construction in everything that is processed.

The Bill of Material, which is the result of the estimator's takeoff, is a list of the materials in the installation standard drawings. This document is used by the purchasing agent to purchase the materials required for the job. This document reflects the stages of construction and can be used by the job management supervisor for overall scheduling of material and tools. Fortunately for the electrical contracting industry, the contractor determines the type of material used in the installation of branch circuit wiring and most other installation methods and procedures; thus the procedures referred to are controlled almost entirely by the contractor, and their development may be done with the assurance that the procedures are acceptable. The drawings may be submitted to the architect for approval before the installation is scheduled.

From the moment the type of installation is selected in accordance with the type of construction, through the purchasing and warehousing stages, the installation standards provide the means of communication between the internal channels of the business and the field. Thus, the installation standard drawings are a most valuable document, used to inform specialists within the business of the processes which have preceded them, making possible the performance of their responsibilities in the proper sequence of events.

It is the contractor's responsibility to provide maximum support for the field. No matter how conscientious the field personnel may be, there are many things beyond their control which lie entirely in the domain of the contractor. Successful execution of the work must be done on a team basis. It is also vital that those in the field know the overall requirements of the job, including the initial planning and scheduling, and all subsequent revisions to the schedule.

In order for an electrical contracting business to succeed, it must be composed of individuals who have been properly trained, and between whom communication has been provided such that one specialist can pick up where another leaves off.

A tremendous number of drawings would be required to illustrate *all* the conditions in which the field is involved in the business. However, a number of *typical* drawings have been prepared to illustrate the material assemblies which prevail for certain types of construction. These are shown, courtesy of the Estimate Corporation, on the following pages.

BILL OF MATERIAL FOR TYPICAL ASSEMBLY

KEY	QUAN.	DESCRIPTION	KEY	QUAN.	DESCRIPTION
A	.10	BOLT TOG MH 1/4 X 3	C	1	COND GRC 3
B	.10	WASHER FLAT 1/4	D	.10	STRAP E/G 1H 3

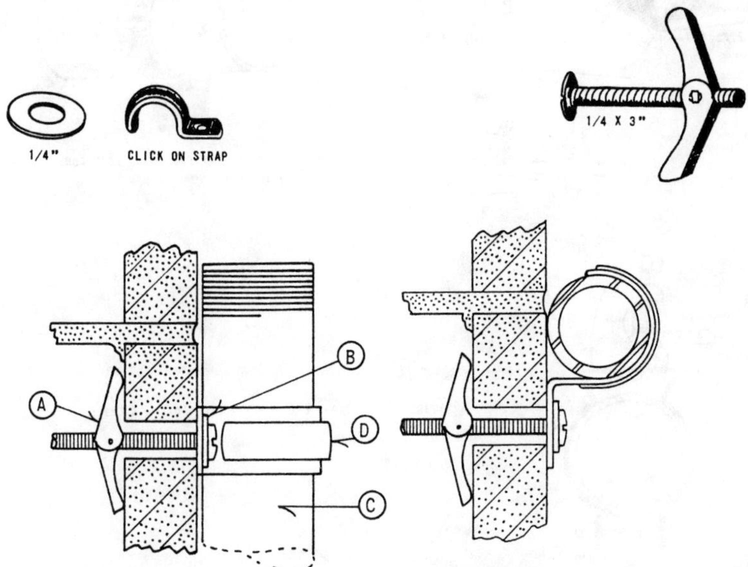

THIS IS THE TYPE OF FASTENING USED FOR:
2-1/2" & 3" CONDUIT.

NOTE: BASIC UNIT OF QUANTITY IS ONE FOOT.

ASSEMBLY STANDARD FOR CONDUIT, GRC EXPOSED ON MASONRY

BILL OF MATERIAL FOR TYPICAL ASSEMBLY

KEY	QUAN.	DESCRIPTION	KEY	QUAN.	DESCRIPTION
A	.10	SCR MR 1/4-20 x 5/8	D	1	COND GRC 2 1/2
B	.10	NUT HEX 1/4-20	E	.10	HGR COND MINR 2 1/2
C	.10	CLAMP BEAM CAST 1/4			

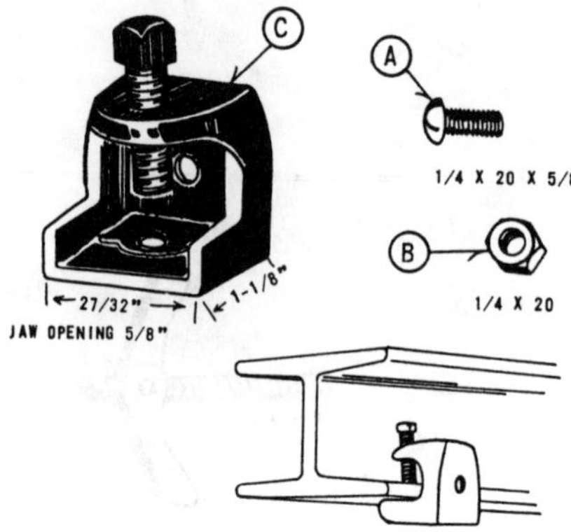

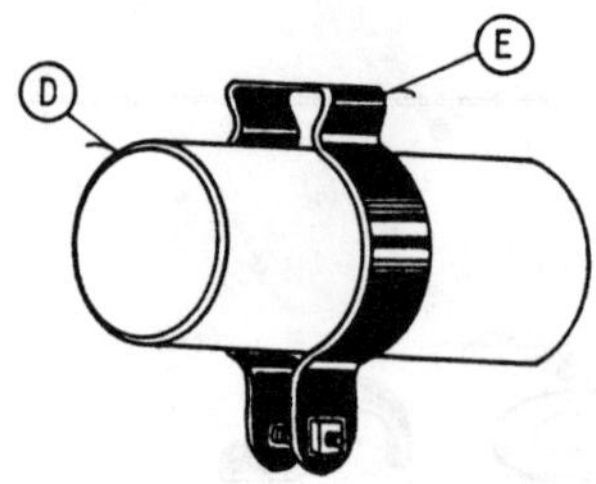

THIS TYPE OF FASTENING USED FOR:
2-1/2" AND 3"

TYPICAL FOR: CXA - CXE - CXF - CXG - CXL - & CXP

NOTE: BASIC UNIT OF QUANTITY IS ONE FOOT.

ASSEMBLY STANDARD FOR CONDUIT, GRC EXPOSED ON STEEL

BILL OF MATERIAL FOR TYPICAL ASSEMBLY

KEY	QUAN.	DESCRIPTION	KEY	QUAN.	DESCRIPTION
A	.10	SCR LAG 3/8 X 2-1/2	C	1	COND GRC 5
B	.10	WASHER FLAT 3/8	D	.10	STRAP GRC 1H CAST 5

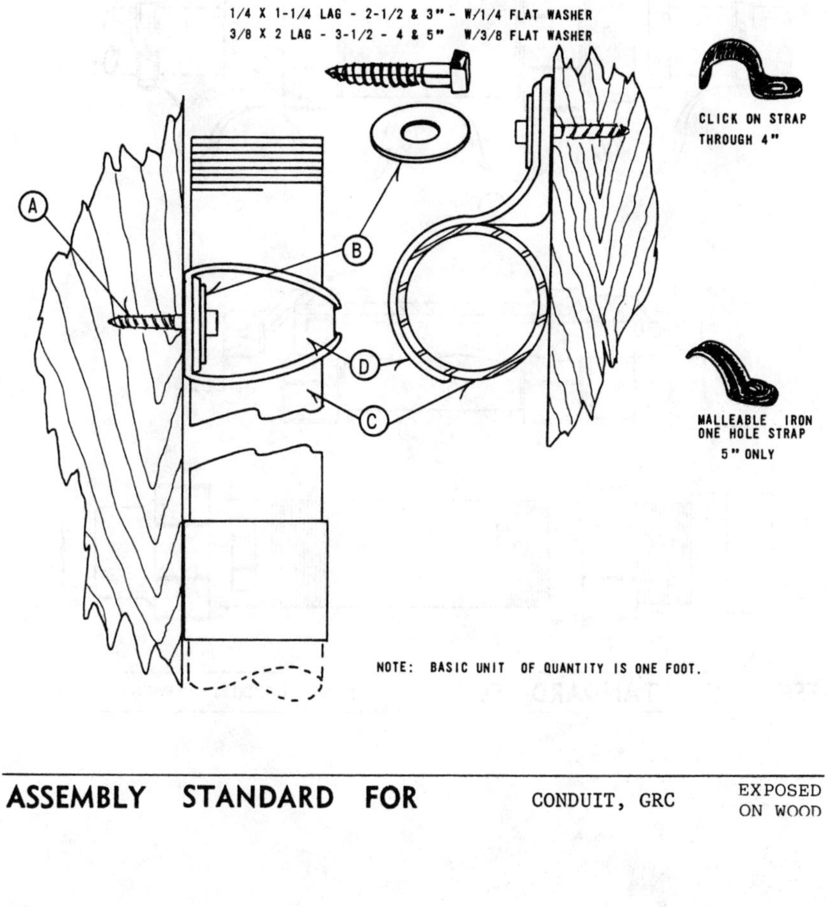

ASSEMBLY STANDARD FOR CONDUIT, GRC EXPOSED ON WOOD

BILL OF MATERIAL FOR TYPICAL ASSEMBLY

KEY	QUAN.	DESCRIPTION	KEY	QUAN.	DESCRIPTION
A	1	EXP JOINT 1 GRC	B	1	JUMPER BONDING 1 IN

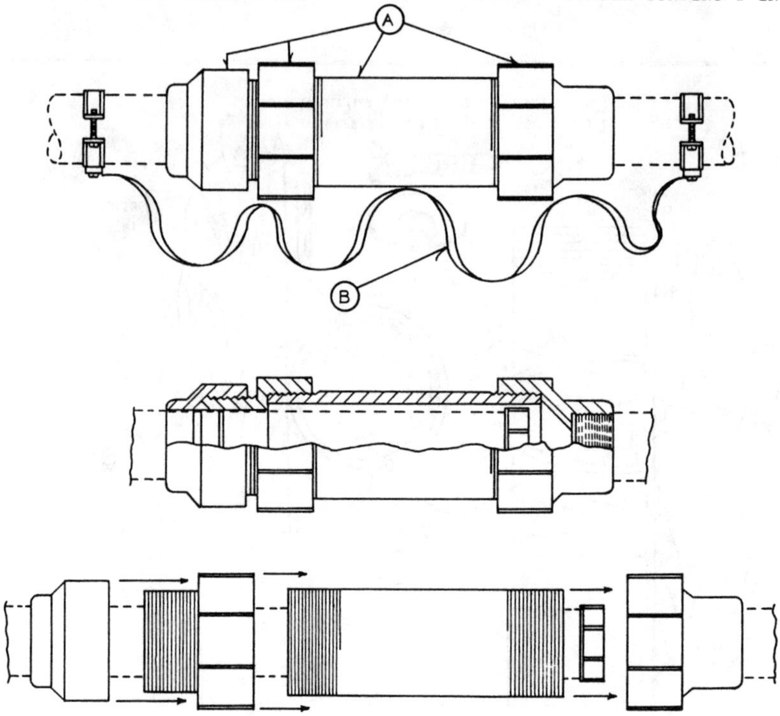

ASSEMBLY STANDARD FOR COUPLING, CONDUIT-EXPANSION

BILL OF MATERIAL FOR TYPICAL ASSEMBLY

KEY	QUAN.	DESCRIPTION
A	2	DUCT UF 1X3
B	.2	FTG BLNK END 1X3
C	.2	SADDLE 2 DUCT 1X3

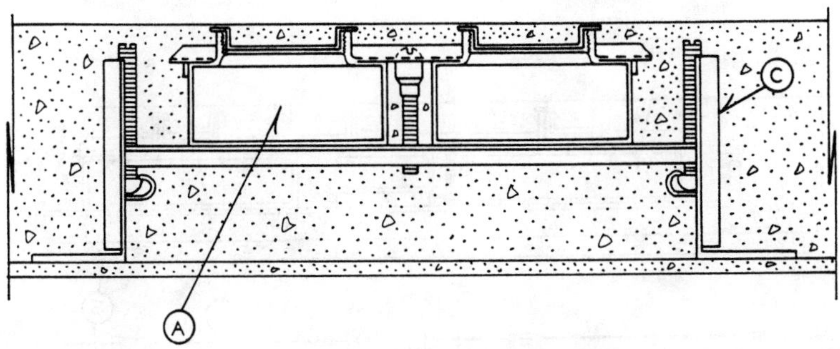

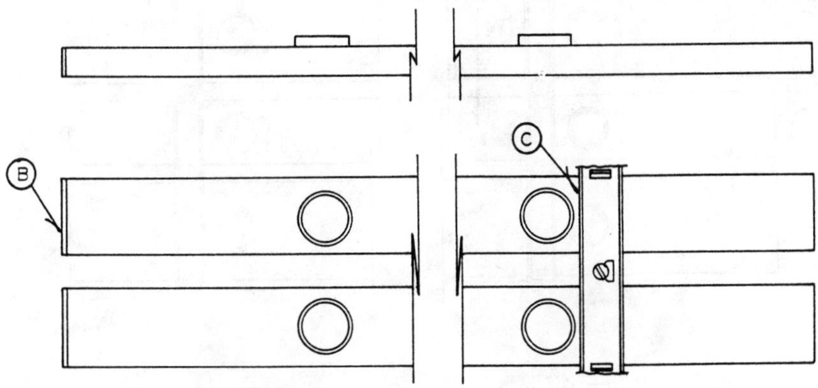

ASSEMBLY STANDARD FOR DUCT, UNDERFLOOR | NO. 2 DUCT

BILL OF MATERIAL FOR TYPICAL ASSEMBLY

KEY	QUAN.	DESCRIPTION
A	1	DUCT UF 1X3
B	2	DUCT UF 1X6
C	.1	COUP UD 1X3
D	.2	COUP UD 1X6
E	.2	SADDLE DUCT 1-2 2-4

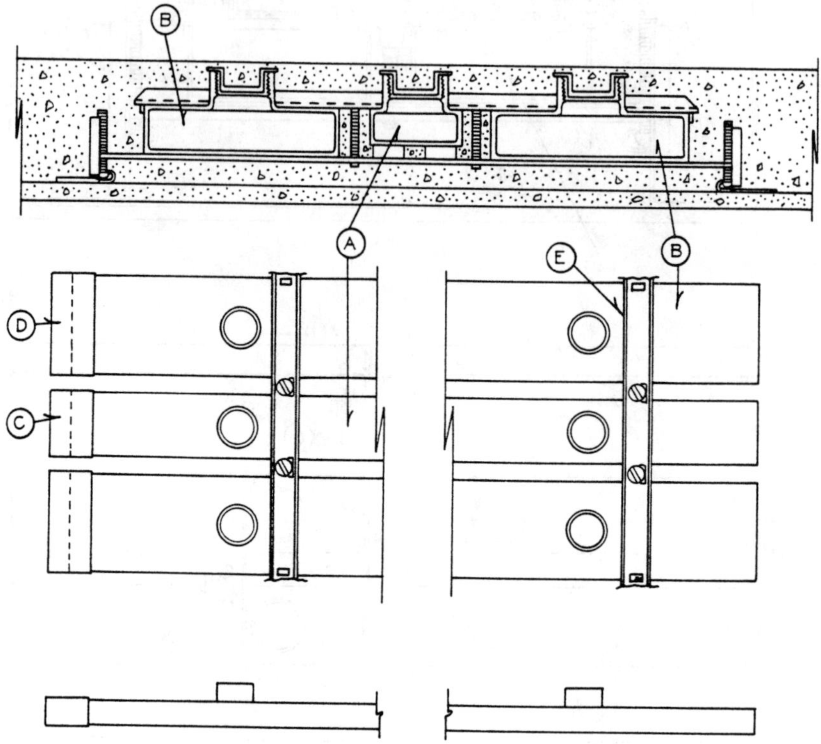

ASSEMBLY STANDARD FOR DUCT, UNDERFLOOR NO. 2 DUCT | NO. 4 DUCT

BILL OF MATERIAL FOR TYPICAL ASSEMBLY

KEY	QUAN.	DESCRIPTION	KEY	QUAN.	DESCRIPTION
A	2	NUT HEX 3/8-16	D	2	ROD ALL THREAD 3/8
B	2	WASHER SQ 3/8	E	2	CHANNEL 1 5/8SQ 12G
C	2	ANCH SELF-DRILL 3/8	F	2	NUT CHANL W/SPG 3/8

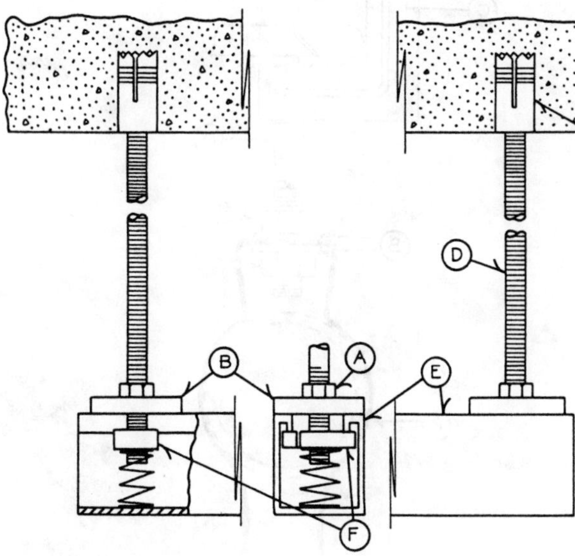

ASSEMBLY STANDARD FOR HANGER, CONDUIT /CHANNEL

BILL OF MATERIAL FOR TYPICAL ASSEMBLY

KEY	QUAN.	DESCRIPTION
A	2	NUT HEX 1/4-20
B	2	WASHER FLAT 1/4
C	1	CLAMP BEAM SP STL
D	1	ROD ALL THREAD 1/4
E	1	HGR 1 1/4 EMT/GRC

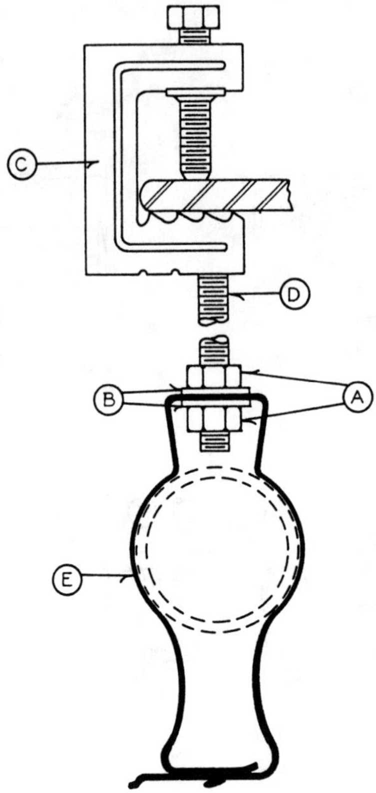

ASSEMBLY STANDARD FOR HANGER, CONDUIT ON STEEL

BILL OF MATERIAL FOR TYPICAL ASSEMBLY

KEY	QUAN.	DESCRIPTION
A	6	WIRE AF 14
B	2	CONN SCOTCHLOK18/12
C	2	EXT FIX 3/8X1 3/8
D	1	LOCKNUT GRC 3/8
E	1	HGR FIX ADJ STEM 22

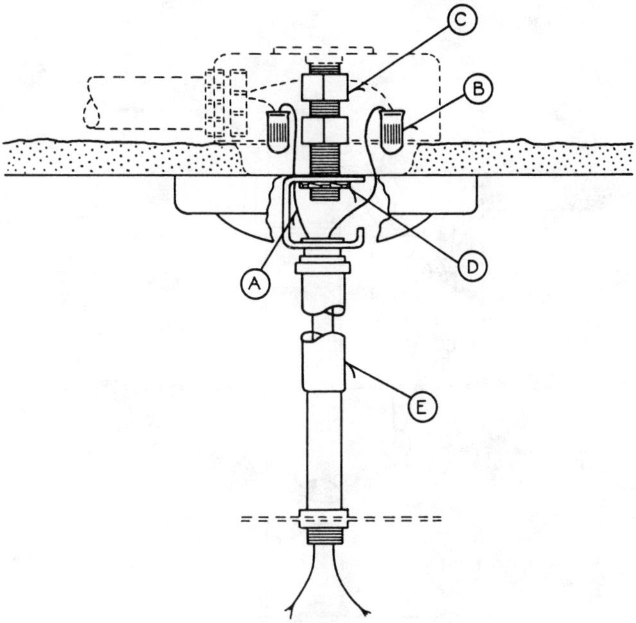

ASSEMBLY STANDARD FOR HANGER, FIXTURE OUTLET MTD

BILL OF MATERIAL FOR TYPICAL ASSEMBLY

KEY	QUAN.	DESCRIPTION	KEY	QUAN.	DESCRIPTION
A	4	WIRE AF 14	E	2	BUSH GRC 1/2
B	2	CONN SCOTCHLOK18/12	F	2	LOCKNUT GRC 1/2
C	1	COV 40CT FLT 1/2KO	G	1	NIP GRC 1/2 X 12
D	1	HICKEY FIX 3/8X1/2			

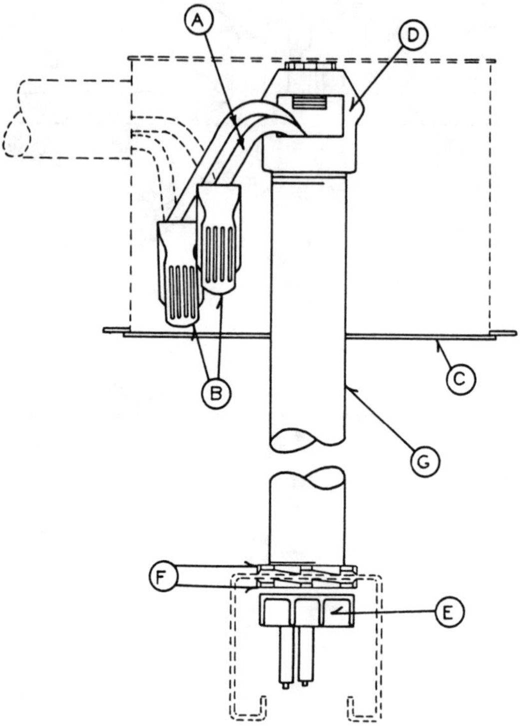

ASSEMBLY STANDARD FOR HANGER, FIXTURE OUTLET BOX MTG

BILL OF MATERIAL FOR TYPICAL ASSEMBLY

KEY	QUAN.	DESCRIPTION	KEY	QUAN.	DESCRIPTION
A	4	WIRE AF 14	D	1	LOCKNUT GRC 3/8
B	2	CONN SCOTCHLOK18/12	E	1	HGR FIX TWIN STEM 8
C	2	EXT FIX 3/8X1 3/8			

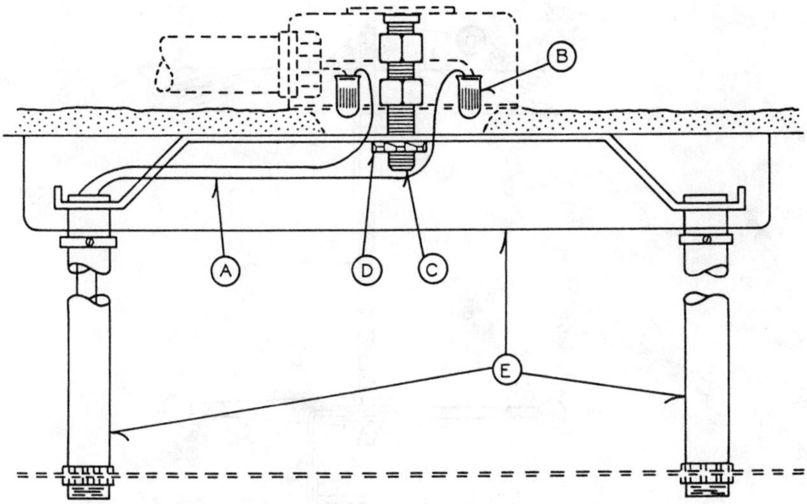

ASSEMBLY STANDARD FOR HANGER, FIXTURE TWIN

BILL OF MATERIAL FOR TYPICAL ASSEMBLY

KEY	QUAN.	DESCRIPTION	KEY	QUAN.	DESCRIPTION
A	2	NUT HEX 1/4-20	D	1	SCR COACH 1/4 3
B	2	WASHER FENDER 1/4	E	5	ROD ALL THREAD 1/4
C	1	COUP ROD 1/4	F	1	FTG CHNL UNIS 2335

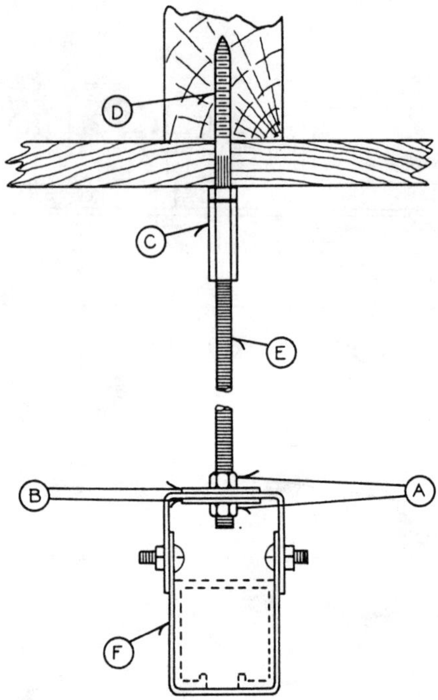

ASSEMBLY STANDARD FOR HANGER, FIXTURE | EXP.ON WOOD

BILL OF MATERIAL FOR TYPICAL ASSEMBLY

KEY	QUAN.	DESCRIPTION
A	2	NAIL ROOF 1 1/2
B	2	CONN SCOTCHLOK16/10
C	1	RNG 4CONC 3 COMB
D	1	PL 4CONC RNG W/STUD
E	2	BUSH GRC 1/2 CAP
F	1	WASH REDR 3/4KO1/2
G	4	LOCKNUT GRC 1/2

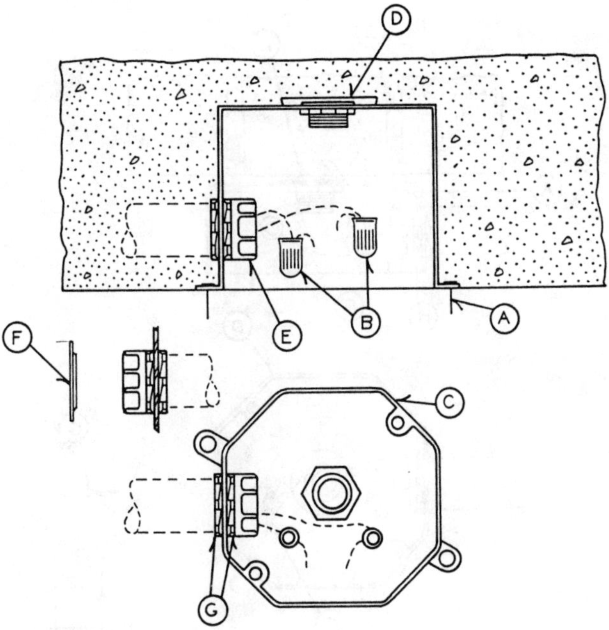

ASSEMBLY STANDARD FOR OUTLET, FIXTURE | DECK FLUSH

BILL OF MATERIAL FOR TYPICAL ASSEMBLY

KEY	QUAN.	DESCRIPTION	KEY	QUAN.	DESCRIPTION
A	2	SCR SMTL P 10 3/4	E	1	STUD FIX NOBOLT 3/8
B	2	ANCH PLASTIC 10 1	F	2	BUSH GRC,1/2
C	2	CONN SCOTCHLOK16/10	G	1	WASH REDR 3/4KO1/2
D	1	BOX 4OCT 1 1/2 COMB	H	4	LOCKNUT GRC 1/2

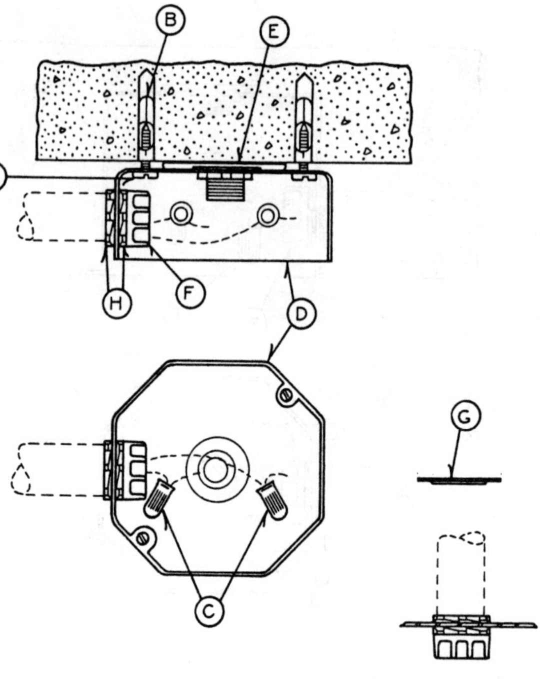

ASSEMBLY STANDARD FOR OUTLET, FIXTURE EXP.ON CONC.

BILL OF MATERIAL FOR TYPICAL ASSEMBLY

KEY	QUAN.	DESCRIPTION	KEY	QUAN.	DESCRIPTION
A	2	BOLT TOGGLE 3/16 4	E	2	BUSH GRC 1/2
B	2	CONN SCOTCHLOK16/10	F	1	WASH REDR 3/4KO1/2
C	1	BOX 4OCT 1 1/2 COMB	G	4	LOCKNUT GRC 1/2
D	1	STUD FIX NOBOLT 3/8			

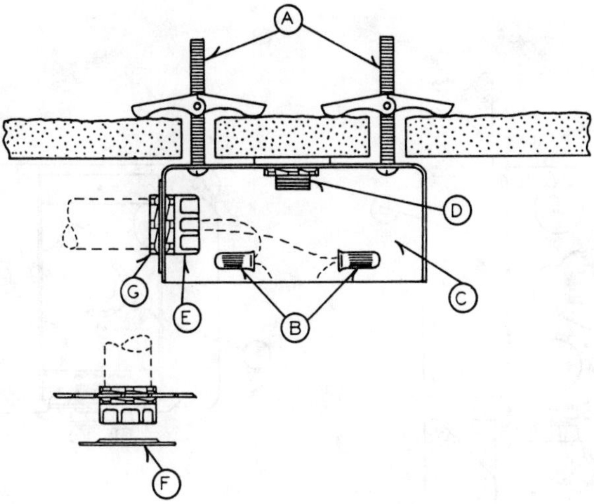

ASSEMBLY STANDARD FOR OUTLET, FIXTURE-exposed on ceil.

BILL OF MATERIAL FOR TYPICAL ASSEMBLY

KEY	QUAN.	DESCRIPTION
A	2	CONN SCOTCHLOK16/10
B	1	COV 4SQ 1 DEV 1 1/2
C	1	BOX 4SQ 1 1/2 1/2
D	1	STUD FIX NOBOLT 3/8

KEY	QUAN.	DESCRIPTION
E	2	PUSH PENNY PLAS 1/2
F	2	CONN EMT 1/2
G	1	CONST LABOR BOX

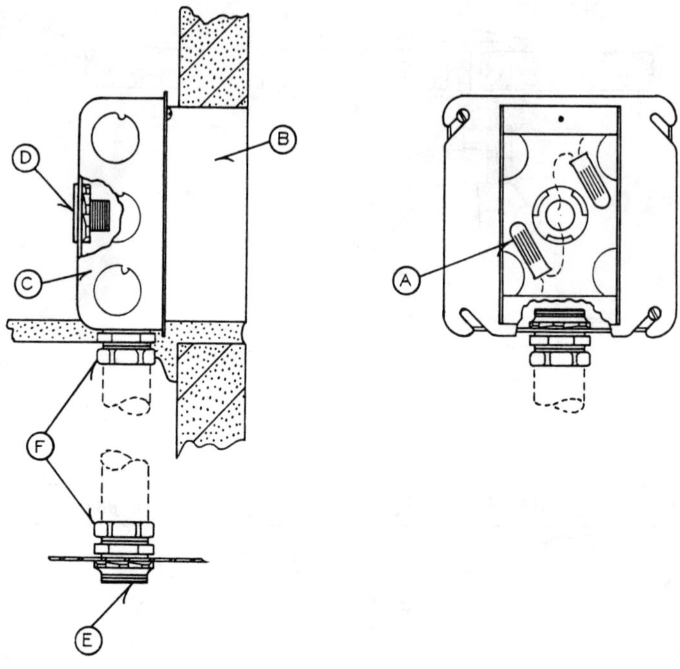

ASSEMBLY STANDARD FOR OUTLET, FIXTURE MASONRY FLUSH

BILL OF MATERIAL FOR TYPICAL ASSEMBLY

KEY	QUAN.	DESCRIPTION	KEY	QUAN.	DESCRIPTION
A	1	ASSEMBLY BOX FORM	D	1	BOX 4SQ 1 1/2 1/2
B	2	CONN SCOTCHLOK16/10	E	2	BUSH GRC 1/2 CAP
C	1	COV 4SQ 1DEV 3/4	F	4	LOCKNUT GRC 1/2

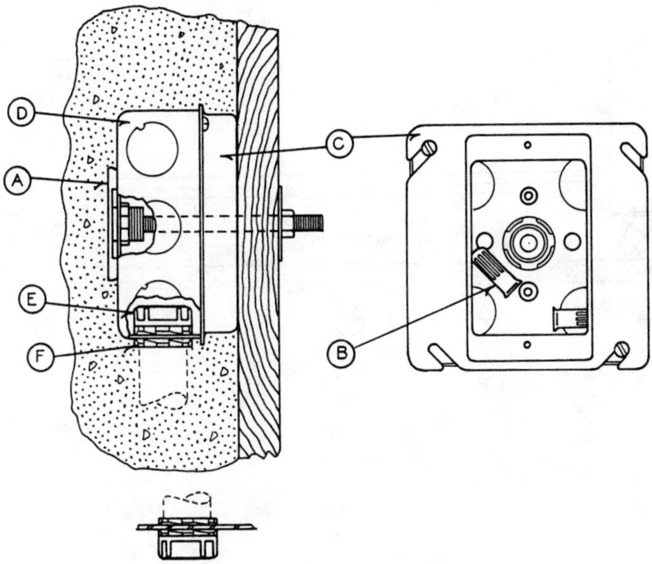

ASSEMBLY STANDARD FOR	OUTLET, FIXTURE	REINF. CONC. FLUSH

BILL OF MATERIAL FOR TYPICAL ASSEMBLY

KEY	QUAN.	DESCRIPTION	KEY	QUAN.	DESCRIPTION
A	4	NAIL ROOF 1 1/2	D	1	COV 4OCT-3RND 5/8
B	2	CONN SCOTCHLOK 16/10	E	1	BOX 4OCT 1 1/2 RMX
C	1	COV 3 OCT FLAT BLNK	F	1	HGR BAR 18 ADJ STUD

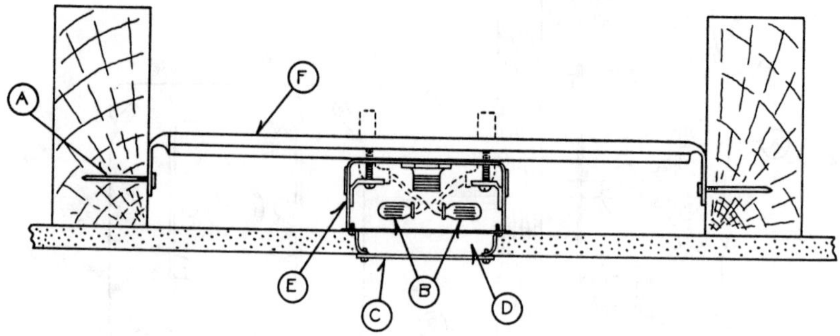

ASSEMBLY STANDARD FOR OUTLET, JUNCTION-WOOD FRAME, FLUSH

BILL OF MATERIAL FOR TYPICAL ASSEMBLY

KEY	QUAN.	DESCRIPTION	KEY	QUAN.	DESCRIPTION
A	2	NAIL ROOF 1 1/2	E	1	WASH REDR 1 KO 3/4
B	1	COV 4SQ 1DEV 1/2	F	2	PUSH PENNY PLAS 1
C	1	BOX 4SQ 2 1/8 1	G	2	CONN EMT 1
D	1	HGR 4SQ BOX	H	1	PL TEL 1G IV MTL

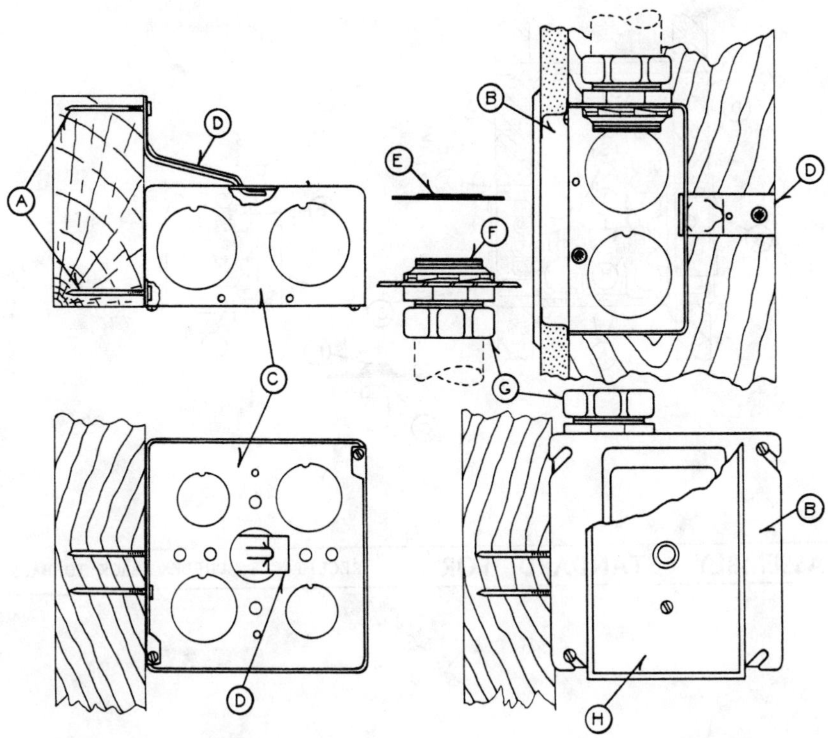

ASSEMBLY STANDARD FOR OUTLET, TELEPHONE WOOD FRAME FLUSH

BILL OF MATERIAL FOR TYPICAL ASSEMBLY

KEY	QUAN.	DESCRIPTION
A	2	PIGTAIL RECP GNDNG
B	2	RECP DUP 15 STD IV
C	2	COV 4SQ 1DEV 1 1/2
D	2	BOX 4SQ 1 1/2 CMB
E	2	BUSH GRC 1/2
F	2	PUSH PENNY PLAS 3/4
G	2	CONN EMT 3/4
H	4	LOCKNUT GRC 1/2
I	1	NIP GRC 1/2 X 2 1/2
J	2	PL DR 1G IV MTL

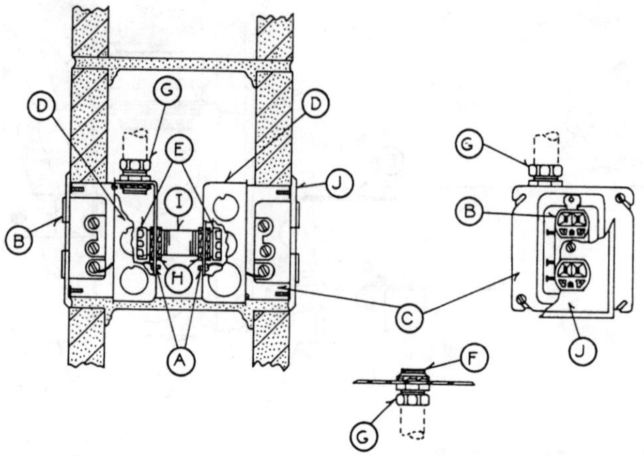

ASSEMBLY STANDARD FOR RECEPTACLE, DUPLEX | BACK TO BACK

BILL OF MATERIAL FOR TYPICAL ASSEMBLY

KEY	QUAN.	DESCRIPTION	KEY	QUAN.	DESCRIPTION
A	1	SCR SMTL P 10 3/4	D	1	COV IND 4SQ 1DR
B	1	ANCH PLASTIC 10 1	E	1	BOX 4SQ 1 1/2 1/2
C	1	RECP DUP 15 DLX BRN	F	2	CONN EMT 1/2

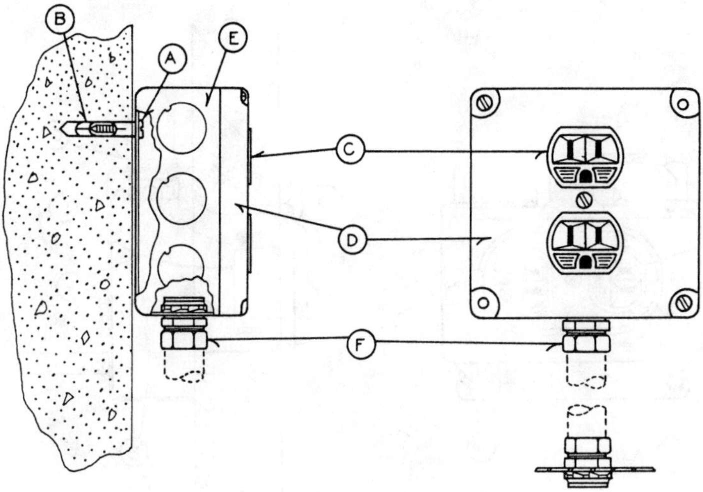

ASSEMBLY STANDARD FOR RECEPTACLE, DUPLEX -EXP. ON CONC

BILL OF MATERIAL FOR TYPICAL ASSEMBLY

KEY	QUAN.	DESCRIPTION	KEY	QUAN.	DESCRIPTION
A	1	PIGTAIL RECP GNDNG	E	1	BOX 4SQ 1 1/2 1/2
B	1	COV DR 1G WP	F	2	BUSH GRC 1/2 CAP
C	1	RECP DUP 15 DLX BRN	G	4	LOCKNUT GRC 1/2
D	1	COV 4SQ 1 DEV 1 1/2			

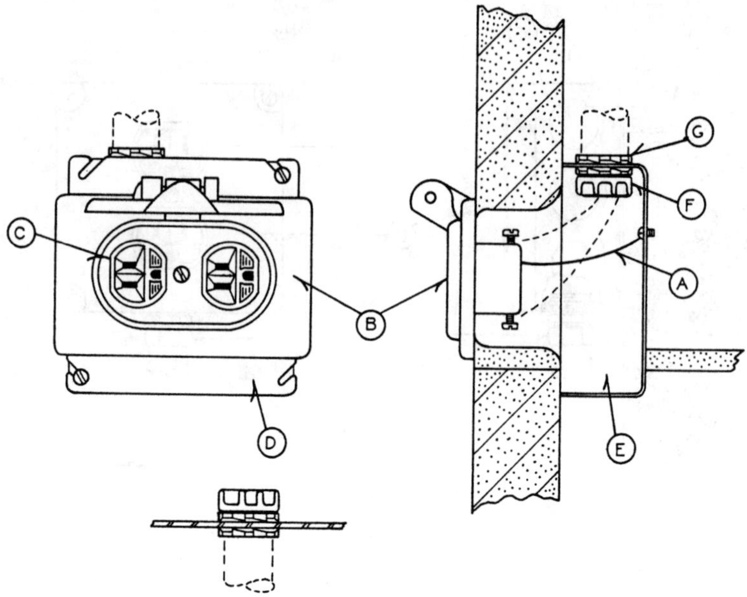

ASSEMBLY STANDARD FOR RECEPTACLE, DUPLEX-FLUSH MASON"Y

BILL OF MATERIAL FOR TYPICAL ASSEMBLY

KEY	QUAN.	DESCRIPTION
A	1	START MAG 230V 3P NEMA 1 SIZE 2
B	3	WASHER FLAT
C	3	BOLT TOG MH 1/4X3
D	12	WIRE CU 6 THW
E	4	WIRE CU 8 TW GRN
F	1	LUG TERM CU AL
G	3	CONN SCOTCHLOK 12/6
H	3	COND WP FLX 1
I	1	CONN FLXWP ST 1
J	1	CONN FLXWP 90 1
K	1	COUP GRC 1

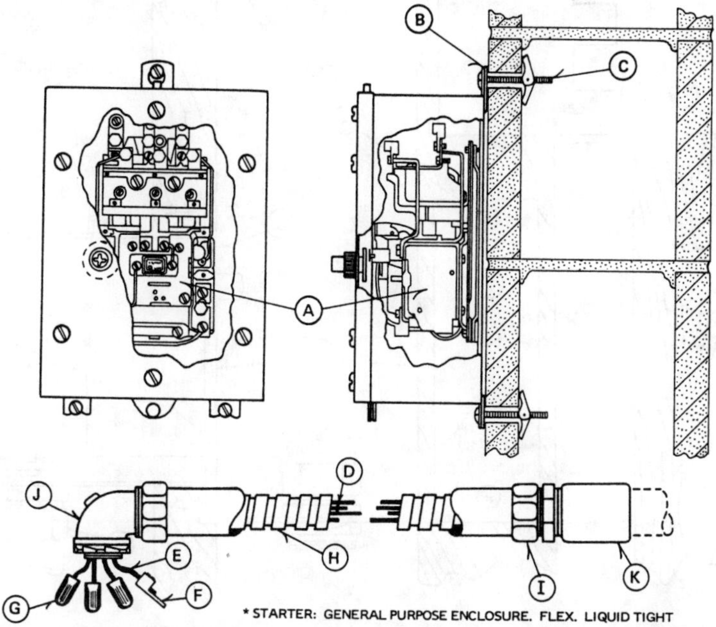

* STARTER: GENERAL PURPOSE ENCLOSURE. FLEX. LIQUID TIGHT

ASSEMBLY STANDARD FOR STARTER, MAGNETIC | EXP. ON HOLLOW MASONRY

BILL OF MATERIAL FOR TYPICAL ASSEMBLY

KEY	QUAN.	DESCRIPTION	KEY	QUAN.	DESCRIPTION
A	2	SCR WD FH NO6 3/4	C	1	BOX DOOR SWITCH
B	2	CONN EMT 1/2	D	1	SW DOOR NORM OPEN

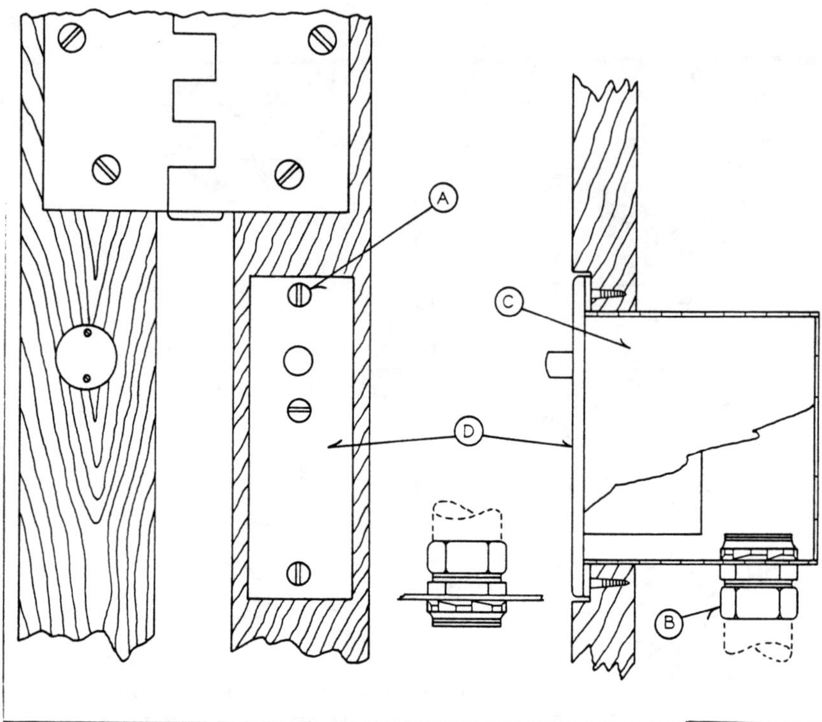

ASSEMBLY STANDARD FOR SWITCH, DOOR | WOOD FRAME FLUSH

BILL OF MATERIAL FOR TYPICAL ASSEMBLY

KEY	QUAN.	DESCRIPTION	KEY	QUAN.	DESCRIPTION
A	4	SCR SMTL P 10 1-1/4	D	1	SWSAF 3P+N 600HG 60
B	4	ANCH PLASTIC 10 1	E	8	TERM LABOR 06 WIRE
C	3	FUSE NON 600V 60A			

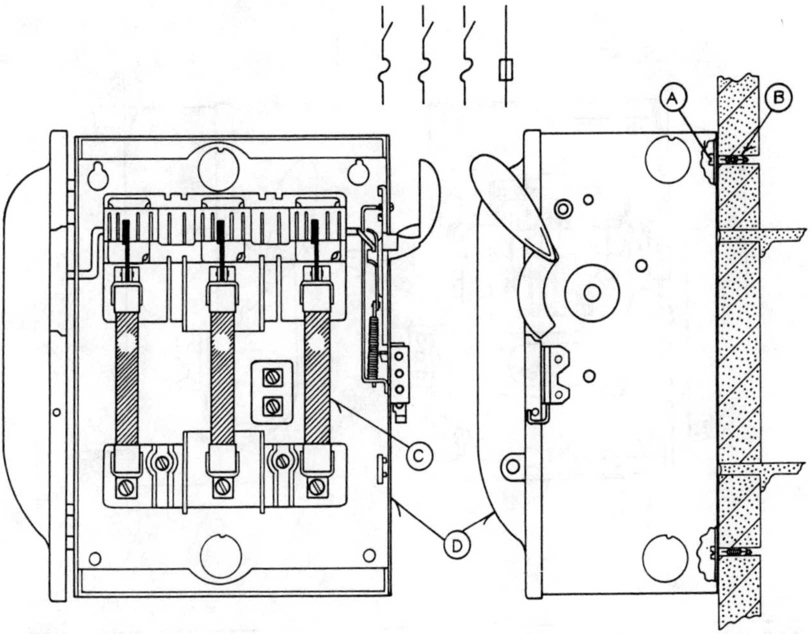

ASSEMBLY STANDARD FOR SWITCH, FUSIBLE SAFETY EXP. ON HOLLOW MASONRY

BILL OF MATERIAL FOR TYPICAL ASSEMBLY

KEY	QUAN.	DESCRIPTION	KEY	QUAN.	DESCRIPTION
A	4	SCR SMTL P 10 1-1/4	C	1	SWSAF 3PNF 600HG 30
B	4	ANCH PLASTIC 10 1	D	6	TERM LABOR 10 WIRE

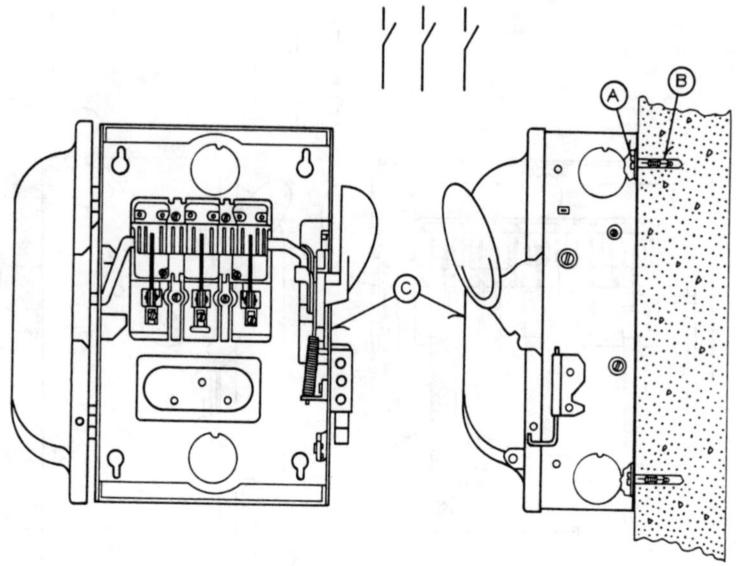

ASSEMBLY STANDARD FOR SWITCH, NON-FUSIBLE SAFETY-ON CONC

BILL OF MATERIAL FOR TYPICAL ASSEMBLY

KEY	QUAN.	DESCRIPTION	KEY	QUAN.	DESCRIPTION
A	2	ASSEMBLY BOX FORM	D	4	LOCKNUT GRC 3/4
B	1	BOX MSNRY 3 1/2 4G	E	4	SW TUM 10 1P STD BR
C	2	BUSH GRC 3/4 CAP	F	1	PL SW 4G 040 BRS

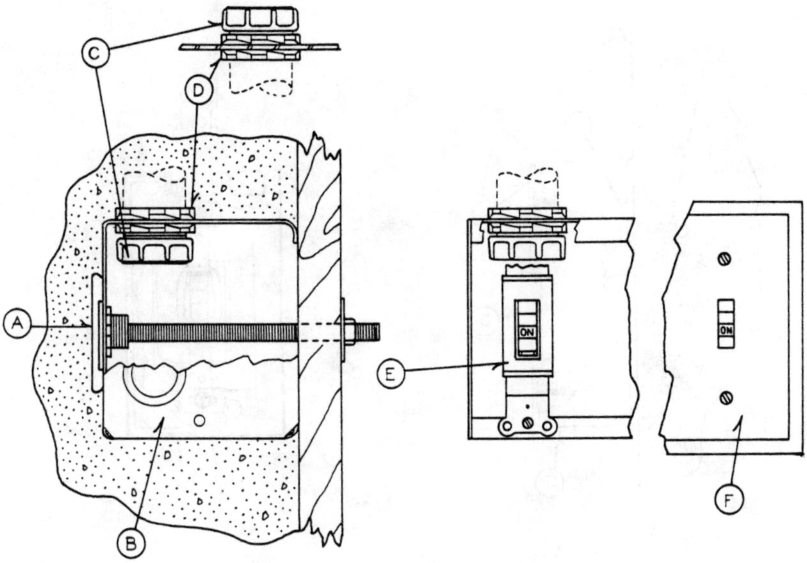

ASSEMBLY STANDARD FOR SWITCH, TUMBLER-FLUSH CONC.

BILL OF MATERIAL FOR TYPICAL ASSEMBLY

KEY	QUAN.	DESCRIPTION	KEY	QUAN.	DESCRIPTION
A	1	SCR SMTL P 10 3/4	D	1	SW TUM 20 1P STD IV
B	1	ANCH PLASTIC 10 1	E	1	PL SW 1G IV MTL
C	1	BOX DEV WMLD 5748			

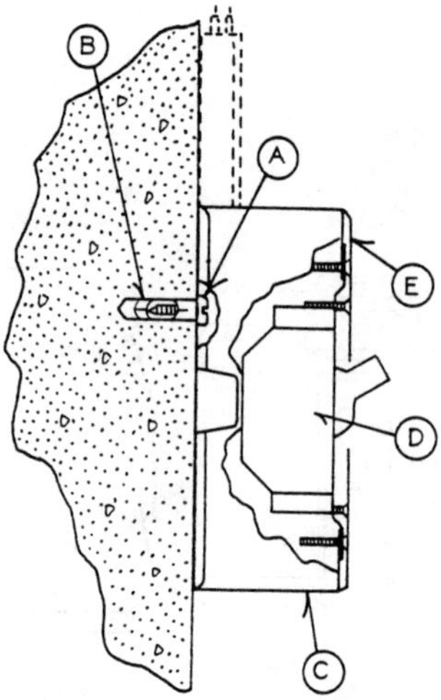

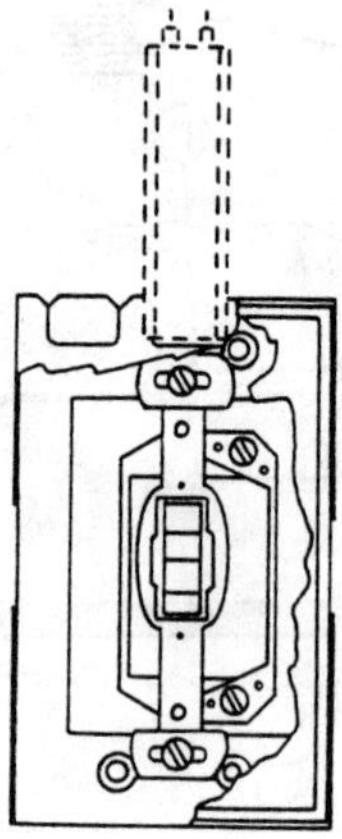

ASSEMBLY STANDARD FOR SWITCH, TUMBLER SURFACE (WMLD)

BILL OF MATERIAL FOR TYPICAL ASSEMBLY

KEY	QUAN.	DESCRIPTION	KEY	QUAN.	DESCRIPTION
A	.10	SCR CAP 3/8 X 1-1/2	D	1	COND GRC 5
B	.10	WASHER FLAT 3/8	E	.10	STRAP GRC 1H CAST 5
C	.10	ANCH SELF-DRILL 3/8			

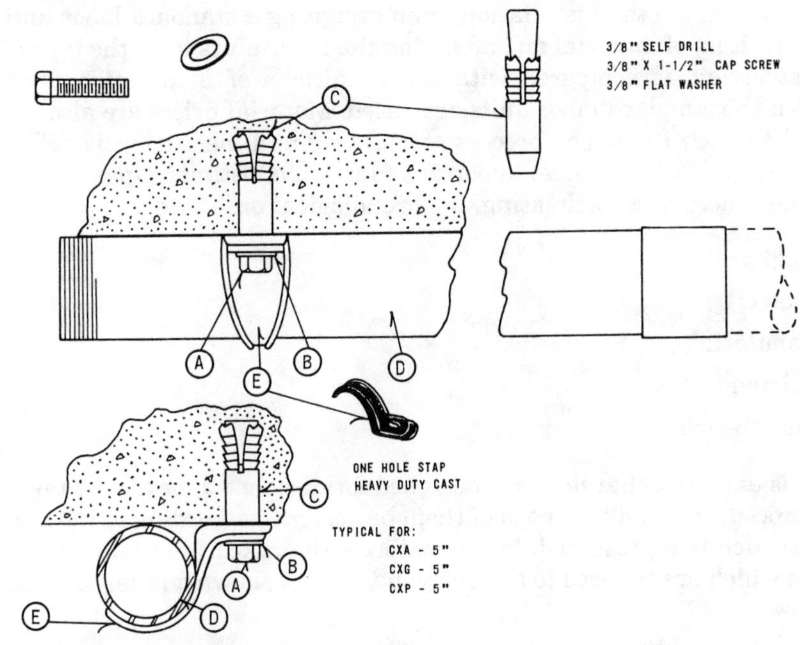

NOTE: BASIC UNIT OF QUANTITY IS ONE FOOT.

ASSEMBLY STANDARD FOR CONDUIT, GRC EXPOSED ON CONCRETE

Principle 5: Take Off the Estimate by Systems

As we have said before, the estimator forecasts costs of installing electrical work consisting of material, labor, and direct job expense by envisioning from plans and specifications the conditions under which the installation will be made. The process of envisioning is one of counting, measuring, and listing the quantities of items of material required to make the installation, then assigning a standard labor unit to each item of material and adjusting the labor units to fit the type of construction as compared with the conditions of installation upon which the standard labor units are based. Material prices are also applied to each item. The process of counting and measuring is called *takeoff,* and it is the most important function of estimating.

The functions of estimating, in chronological order, are:

- Registration
- Takeoff
- Tabulating
- Pricing
- Summarizing

It is essential that properly designed forms be used for all the takeoff procedures, and that each of them be designed according to the system which is represented. In other words, the takeoff is made by systems which are tailored to the divisions of work shown on the plans, as follows:

- Service, metering, and grounding
- Lighting fixtures, lamps, and hangers
- Distribution equipment, switches, and breakers
- Raceway equipment and supports
- Feeder and busway equipment and auxiliaries
- Branch circuit conduit and wire
- Ceiling, switch, receptacle, and other outlets
- Power system conduit, wire, and outlets
- Telephone system conduit and outlets
- Special systems conduit, wire, outlets, and equipment

Before an estimator begins the takeoff it is mandatory that the electrical work be completely defined in such a manner that the counting and measuring process can proceed. If the plans are not designed in a manner that makes this possible, it will be necessary for the estimator to so define them. The estimator will need to provide a complete definition of elements of the electrical system, as follows:

- Prepare a one-line diagram of the electrical system
- Lay out the routing of feeders and determine the sizes
- Lay out the branch circuit system and determine conduit sizes and number of conductors
- Determine the size, number, and location of panelboards
- Check the architectural plans and determine the types of construction for ceilings, walls, and floors as well as elevations and ceiling heights

In order to ensure accurate application of labor units, it is essential to identify the types of construction and the conditions of installation for every operation involved and to associate each of them properly with the items of material to be installed. It is here that standards become extremely important, because if conditions of installation and tooling are carefully defined for each labor unit, the estimator must be careful to identify the exceptions to the standards—and to accomplish this the estimator must thoroughly understand the standards. It is also necessary for the estimator to compare accurately the actual conditions of the job at hand with the standard conditions of installation and to be able to compensate for the differences by applying the appropriate adjustments for height of installation, density of work, length of pull, inadequate tooling, or any nonstandard working conditions.

Order of takeoff

The time allowed by an owner or architect for bidding a job is seldom adequate, so the estimator is usually rushed to complete the estimate. Many things can be done by members of the contractor's organization to provide maximum support to the estimator by way of securing material and equipment prices, adding and extending worksheets, and securing information from the architect or engineer. In turn, the estimator must schedule the order of takeoff to make available to the support functions the lists of material and equipment which require price quotations and clerical processing. The items which usually require such information are lighting fixtures, panelboards, and service equipment such as underfloor duct.

The following order of takeoff must be observed:

- Lighting fixtures, hangers, and lamps
- Service, metering, and grounding

- Switchboards, panelboards, and breakers
- Underfloor duct and busway systems
- Special systems

Forms required for takeoff

Many forms are required for the preparation of an estimate. Each form is designed to accomplish a specific phase of the work, and each enables the estimator to record data, tabulate data, or list for pricing and laboring.

Not only do printed estimating forms provide the format for properly organizing each of the phases of the estimate, but they also ensure accuracy and prevent oversights and omissions by indicating elements of work to be recorded, and conditions of installation to be recognized and indicated on the form for future use. In addition, they act as work organizers which outline consecutive steps for accomplishing the complete task. The use of printed forms can do a great deal to increase the productivity of an estimator and ensure the accuracy of the estimates.

The forms used in takeoff are the following:

Form no.	Form title
E-3	Service, Metering, and Grounding Takeoff Sheet
E-4	Lighting Fixture Takeoff Schedule
E-5	Distribution Equipment Takeoff Schedule
E-6	Raceway Equipment Takeoff Schedule
E-7	Feeder and Busway Takeoff Schedule
E-8	Branch Circuit Takeoff Schedule
E-10	Outlet Detail and Takeoff Schedule
E-11	Power System Takeoff Schedule
E-12	Telephone System Takeoff Schedule
E-13	Special Systems Takeoff Schedule

Each of these forms will be shown, and a description of its main features given.

Form E-3: Service, Metering, and Grounding Takeoff Sheet. Form E-3 is used to record takeoff information for the materials and equipment required for the installation of the main service up to, but not including, the switchboard, as follows:

- Primary supply such as a transformer bank, crib, or vault
- Service conductor installed from the transformer secondary to the switchboard, but not including the switchboard

- Service raceway from the transformer secondary to the switchboard
- Metering equipment, including the current transformers, cabinets, and meter leads
- Grounding materials, including the electrode, clamps, cable, and lugs
- Other equipment such as capacitors, arrestors, and cutouts

Since information relative to the main service does not appear on any other takeoff form, no tabulating is necessary and the information on this form is transferred directly to Form E-19, Pricing Sheet, Equipment Work Schedule and Conductor Work Schedule.

Typical information is shown on a copy of Form E-3 in Fig. 6.3.

Form E-4: Lighting Fixture Takeoff Schedule. This form, shown in Fig. 6.4 with typical information inserted, is used to record all the takeoff information for the lighting system, including fixtures, lamps, hangers, and appurtenances. Since all the takeoff information for the lighting system is shown on this form only, it is not necessary to tabulate the information, and the summarized items are transferred from Form E-4 directly to Form E-19, Pricing Sheet, under Lighting Fixture Work Schedule.

On large jobs it is often desirable, for scheduling purposes, to take off fixtures by the floor, or by the building when separate buildings are involved; in such cases a separate form is required for each subdivision of work. Form E-4 can also be used in the field as a fixture schedule during construction because all the information the electrician requires to make the installation is shown on this form.

During the process of takeoff, the estimator must designate the items, if any, which are installed at a height above 14 ft, as indicated in the top section of the form, for purposes of factoring the standard labor units. This can be accomplished by placing a dot in the square on the form in which the quantity of fixtures appears that are subject to factoring for a height up to 25 ft. An "X" is marked in the square with the quantity which is installed up to 35 ft. Designations made in this way will be used when the information is transferred to Form E-19, Pricing Sheet.

It is essential for the information on the form to be filled in completely at the time of takeoff, as follows:

- It is necessary to provide for each fixture the plan designation, the quantity, a description, the wattage, voltage, manufacturer's name, and catalog number.
- It is necessary to show the type of mounting and the type of construction on which it is mounted.
- It is also necessary to indicate the type of hanger, lamps, aligner, dimmer, and timing device.

Form E-3

Service, Metering, and Grounding

Sheet no. 1 of 1
Estimator RCP
Date 11-6-84

Job name Littleton Arts Center

	Material item	Description	Quantity	Total
Source	Transformer, type and size	By Utility		
	Voltage, primary	By Utility		
	Voltage, secondary	By Utility		
	Arresters	By Utility		
	Cutouts, or switches	By Utility		
	Insulators	By Utility		
	Hardware	By Utility		
	Wire, type and size	By Utility		
Service	Conductor, type and size	500 MCM - THW	68 ft	272 ft
	Conductors, number of	4-2 runs in Parallel		
	Entrance fitting, type and size	A Condulet	2	
	Length of wire pull		26 ft	
	Terminals, wire size	500 MCM Lugs	8	8
		Burndy Connector	8	8
Raceway	Raceway, type and size	3 1/2 in GRS		
	Length of run	2 Runs	26 ft	52 ft
	Type of construction	Exp: on Masonry		
	No. runs in parallel	2		
	Elbows	3 1/2 in - 90°	2	4
	Fittings	A Condulet and 4 Hole Cover	2	4
	Offsets or bends	No		
	Terminals, conduit	4 LN 2 Bushing	2	4
Metering	Housing, type	1 - Type B	1	1
	Cabinet, current trans.	24 in x 36 in CT Cabinet	1	1
	Meter leads – conduit size	1 in EMT	20 ft	20 ft
	wires, size	No 10 THW - 7 Conductor	10 ft	70 ft
Grounding	Clamp, type and size	Clamp - Approved	1	1
	Cable, type and size	0 THW	32 ft	32 ft
	Electrode, type	Water pipe-4 in	1	1
	Terminals, wire	Size 0 - Solderless lug	1	1
	Conduit	3/4 in EMT	30 ft	30 ft
Miscellaneous	Bushing	3/4 in Grounding bushing	1	1

Figure 6.3 Service, Metering, and Grounding Takeoff Sheet.

Provisions are made on the form for each of these features.

Form E-5: Distribution Equipment Takeoff Schedule. Form E-5 is used to record takeoff information relative to the electrical distribution equipment required as shown on the plans or specifications, as follows:

- The main load center, including the main circuit breaker—and including wire terminals and conduit entrances

Form E-4

Lighting Fixture Takeoff Schedule

JOB NAME Littleton Arts Center

SHEET NO 1 OF 1
FLOOR NO 1 BLDG A
ESTIMATOR R.C.P.
DATE 11-9-84

SIGNALS USED ON FORMS FOR FLAGGING DEVIATIONS FROM STANDARDS

HEIGHT OF INSTALLATION		CONDUIT PARALLEL RUNS		WIRE PULLS IN PARALLEL		MISCELLANEOUS SIGNALS
Height	Designation	Number	Designation	Number	Designation	Density of Items — d
15-25 ft	•	2	2r	2	2p	Number of Men — n
26-35-ft	×	3	3r	3	3p	Note: Designation should be made in upper or lower left-hand corner of quantity column square
36-50 ft	✓	4	4r	4	4p	
		5	5r	5	5p	

ITEM NO	PLAN DESIG	QUAN	DESCRIPTION	TYPE FIXTURE	SPEC MFG	MFG CAT NO	FIX. MOUNT	WATTS	VOLTS	HANGER	HGR CONST	LAMPS NO	LAMPS TYPE	LAMPS COLOR	LAMPS WATTS EA	ALIGNER	DIMMER	LOWERING DEVICE	TIMING DEVICE	
1	A	400	Fixture Fl Layin	Layin	Daybe'	16741	T Bar	200	120			4	F-40	WWX	40					
2	B	16	Fixture Inc Recess	Square	Thom	R30C	Rings	150	120			1	A	IF	150					
3	C	22	Fixture VP	Bracket	Appn	72D	Outlet	100	120			1	A	IF	100					
4	D	30	Fixture Fl Strip	Strip	Lith	S130	Drywall	100	120			2	F-40	WWX	40					
5	E	× 44	Fixture RLM Domes	Area	Benj.	V7641	Outlet	100	120	Stem	Outlet	1	Merc Vapor	Color Corr.	100	Box Cover				
6	F	12	Lamp Holders	Porc	A.H&H	140	Outlet	60	120			1	F-60	IF	60					

Figure 6.4 Lighting Fixture Takeoff Schedule.

- All distribution panelboards for lighting and power
- Individual enclosed circuit breakers
- Safety switches, motor starters, and contactors
- Other equipment items such as capacitors

The information developed by the estimator on this form is transferred directly to Form E-19, Pricing Sheet, using the Equipment Work Schedule. In addition to the information entered from this form to the Pricing Sheet there may be other items of equipment from other takeoff forms, but tabulating is not required.

Form E-5 is shown in Fig. 6.5 with some typical takeoff information recorded.

Form E-6: Raceway Equipment Takeoff Schedule. *Raceway equipment* is defined as all types of equipment other than conduit which provide a channel for the installation of wire and cable. Form E-6 is the takeoff form for information pertaining to the raceway, as follows:

- The first section of this form is used to record quantities of Unistrut-type channel and fittings.
- The second section is used to record quantities of cable tray and fittings.
- The third section is used to record quantities of duct and fittings for underfloor duct, header duct, or trench duct.
- The fourth section is used to record quantities of Wiremold-type raceway and fittings.
- The fifth section is used to record quantities of wireway or gutter and fittings.
- The last section is used to record quantities of cable bus or other miscellaneous equipment.

Form E-6 is shown in Fig. 6.6, with typical information inserted.

Form E-7: Feeder and Busway Takeoff Schedule. Form E-7 is used to record the takeoff information relative to the lighting and power feeders—including wire, cable, or busway.

- The first section is devoted to a description of the feeder from its inception to its destination.
- The second section contains information relative to the type and size of raceway.

Form E-5

Distribution Equipment Takeoff Schedule

SHEET NO 1 OF 1
FLOOR NO 1 BLDG A
ESTIMATOR R.C.P.
DATE 11-7-84

JOB NAME Littleton Arts Center

LOADCENTER — TYPE: Freestanding						PANEL BOARDS — TYPE: NQO						MAINS: 3		VOLTS 208		SAFETY SWITCHES — VOLTS					CIRCUIT BREAKERS (ENCL.)					CONTRACTORS (ENCL.)		
VOLTS: 208		MAINS: 800		C.T.'S Cab.		PANEL DESIG	TYPE BRANCH	MOUNT'G	SINGLE-POLE			TWO-POL		THREE-POLE		QUAN	AMPS	TYPE	POLES	MTG	QUAN	AMPS	TYPE	POLES	MTG	QUAN	AMPS	POLES
NO.	TYPE	POLES	AMPS.	W. TERM	C. ENTR					20A																		
1.	ACB	3	1000	8	2-3½																							
2.	MCB	3	400	4	1-2½	LPA	QO	Flush		40																		
3.	MCB	3	225	4	1-2	LPB	QO	Flush		28																		
4.	MCB	3	400	4	1-2½	LPC	QO	Flush		42											2	20	Q	1	Exposed	1	15	1
5.	MCB	3	225	3	1-2	PPA	E	Exposed				2	30	2	100	2	30	C	2	Exposed								
														1	200						1	200	MCB	3	Exposed			
6.	MCB	3	175	3	1-1½	PPB	E	Exposed				3	20	3	60	3	20	C	2	Exposed								
														1	100	1	60	C	3	Exposed								
																1	100	C	3	Exposed						1	100	3
7.	MCB	3	225	3	1-2											1	400	C	3	Exposed						1	START/ 50HP	3

Figure 6.5 Distribution Equipment Takeoff Schedule.

Form E-6

Raceway Equipment Takeoff Schedule

JOB NAME Littleton Arts Center

SHEET NO. 1 OF 1
FLOOR NO. 1 BLDG. A
ESTIMATOR R.C.P.
DATE 11-8-84

CHANNEL (UNISTRUT.)

QUAN	SIZE	TYPE	ANGLE	BAR	BEAM CLAMP QUAN.	BEAM CLAMP SIZE	ROD QUAN.	ROD SIZE	CONDUIT STRAPS QUAN	CONDUIT STRAPS SIZE	SPRING NUTS QUAN.	SPRING NUTS SIZE
30'	P-4000	Hang			58	1/4"	180'	1/4"	16	1½"		
									16	2½"		
100'	P-1000	Mtg	20	5							60	3/8"

CABLE SUPPORT (TRAY)

QUAN.	SIZE	TYPE	COVER	FLAT ELL	VERT ELL	SPLICE PLATE	HANGER

UNDERFLOOR DUCT — HEADER DUCT — TRENCH DUCT

QUAN	TYPE	COMB.	ELL	OFF ELL	VERT. ELL	CAB. CONN.	COND. ADAP. QUAN.	COND. ADAP. SIZE	SERV. FIT QUAN.	SERV. FIT TYPE	JUNC. BOX QUAN.	JUNC. BOX TYPE	SUPPTS
270'	2	2+4	4	1	6	2	16	3/4	36	Elec.	4	224	100
									36	Tel.			
200'	4		4		2	2							

WIREMOLD AND FITTINGS

QUAN.	SIZE	TYPE	OUT BOX	EXT. BOX	JUNC. BOX	FIX. BOX	EXT. ELL	FLAT ELL	INT ELL	END COVER	FITTING	RECEP.	SWITCH

WIREWAY (GUTTER)

NO. PIECES	LENGTH	QUAN.	SIZE	TYPE	COVER	CONN.	END CLO	HANGER	TEE	PAN'L ENTRANCE
2	4'	8'	4"x4"	KO	Screw	1	2			
1	4'	4'	8"x8"	KO	Screw		1			1
4	4'	16'	6"x6"	KO	Screw	4	2	10		

CABLE — BUS

QUAN.	VOLTS	AMPS	ENDS	HANGERS	
20'	5KV	2000	2	4	5FT

Figure 6.6 Raceway Equipment Takeoff Schedule (used for takeoff of channel, tray, duct, mold, and wireway).

- The third section is used to record quantities of raceway or busway fittings.

As indicated before, one of the important functions of a takeoff form is to enable the estimator to indicate with quantities the height of installation, runs in parallel, and runs in the trench. This is done with symbols, or *signals,* which are shown on the form as follows:

	Signal
Height of installation	
15 to 25 ft	•
26 to 35 ft	X
36 to 50 ft	✓
Runs in parallel	r
Pulls in parallel	p
Conduit in trench	t
Riser conduit	R

The total quantities shown on Form E-7 are transferred to Form E-17, Conductor Tabulating Form, along with total quantities of identical items taken from other takeoff forms. The quantities of busway and busway fittings are transferred to Form E-19, Pricing Sheet, under Equipment Work Schedule. The quantities of raceway and the fitting information are transferred to Form E-15, Feeder Conduit Tabulating Form. It is important to recognize that a separate Form E-15 sheet is required for each different type of raceway.

Form E-7 is shown in Fig. 6.7, with a sample of typical information.

Form E-8: Branch Circuit Takeoff Schedule. All branch circuit conduit and wire quantities are recorded on Form E-8 as they are measured by a rotometer or tape. This includes materials for the lighting system, switching, and receptacles, but not the outlets themselves. Nor does it include any wiring materials for the power or telephone systems, including outlets. During the process of counting and measuring, it is important to designate on the form together with the quantity the height of installation, and to indicate properly the type of construction in which the raceway is being installed.

The quantities of conduit and wire summarized on this form are transferred to the following tabulating forms:

- Form E-16, Branch Circuit Tabulating Form
- Form E-17, Conductor Tabulating Sheet

Form E-8 is shown in Fig. 6.8 with typical takeoff information filled in.

Form E-7

Feeder and Busway Takeoff Schedule

JOB NAME *Littleton Arts Center*

SHEET NO. *1* OF *1*
FLOOR NO. *1* BLDG. *A*
ESTIMATOR *R.C.P.*
DATE *11/7/89*

SIGNALS USED ON FORMS FOR FLAGGING DEVIATIONS FROM STANDARDS

HEIGHT OF INSTALLATION		CONDUIT PARALLEL RUNS		WIRE PULLS IN PARALLEL		MISCELLANEOUS SIGNALS
Height	Designation	Number	Designation	Number	Designation	Density of Items — d
		2	2r	2	2p	Number of Men — n
15-25 ft	•	3	3r	3	3p	
26-35 ft	x	4	4r	4	4p	Note: Designation should be made in upper or lower
36-50 ft	✓	5	5r	5	5p	left-hand corner of quantity column square.

FEEDER WIRE, CABLE, OR BUSWAY									RACEWAY								RACEWAY — BUSWAY FITTINGS								
											LENGTH OF RUN														
FEEDER NO.	FROM	TO	DESC.	SIZE	NUMBER OF COND.	LENGTH OF PULL	TYPE	WIRE TERM.	SIZE	TYPE	EMBEDDED	TRENCH	EXPOSED	HANGER	RISER	CONDUIT TERMINAL	ELLS	FITTING	HANGER	END CLOSURE	JUNC. BOXES	TEES	CROSS	C.B. PLUG	V.B. PLUG
2	Switchboard	LPA	Wire	250	4	2@181'	THHN	4	2½"	GRS				x 2@180'		4	2	1-LB	2 Cond. Trap.		1-30"x30"x6"				
3	Switchboard	LPB	Cable	2/0	4	86'	RR	4	2"	GRS		80'				2	2								
4	Switchboard	LPC	Wire	250	4	136'	THHN	4	2½"	GRS	120'					2	2								
5	Switchboard	PPA	Cable	4/0	3	65' 85'	RR	3	2"	GRS	80'	60'				2	1	1-LL			1-16"x16"x4"				
6	Switchboard	PPB	Wire	1/0	3	146'	THHN	3	1½"	GRS				x 2@140'		4	2		2 Cond. Trap.						
7	Switchboard	Air Cond.	Wire	4/0	3	80'	THHN	3	2"	GRS			• 80'			2	2								
8	Switchboard	Shop	Bus	1000A	3	400'	Plug-in	2						400'		1-Bus	4	Cabinet Int.	40	1					10

Figure 6.7 Feeder and Busway Takeoff Schedule.

Form E-8

Branch Circuit Takeoff Schedule

JOB NAME *Littleton Arts Center*

SHEET NO. *1* OF *1*
FLOOR NO. *1* BLDG. *A*
ESTIMATOR *R.C.P.*
DATE *4-7-84*

SIGNALS USED ON FORMS FOR FLAGGING DEVIATIONS FROM STANDARDS

HEIGHT OF INSTALLATION		CONDUIT PARALLEL RUNS		WIRE PULLS IN PARALLEL	
Height	Designation	Number	Designation	Number	Designation
15-25 ft	•	2	2r	2	2p
26-35 ft	x	3	3r	3	3p
36-50 ft	✓	4	4r	4	4p
		5	5r	5	5p

MISCELLANEOUS SIGNALS
Density of Items — d
Number of Men — n
Note: Designation should be made in upper or lower left-hand corner of quantity column square.

CT. NO.	RACEWAY														WIRE — SIZE AND NUMBER OF CONDUCTORS												CABLE W/GROUND WIRE					
		1/2"				1/2"				3/4"						#12				#12				#10								
	TYPE	EXP.	FURRED	FRAME	TERM	EXP.	FURRED	FRAME	TERM	EXP.	FURRED	FRAME	FITT'G	TERM	TYPE	2	3	4	5	2	3	4	5	2	3	4	TYPE	12-2	12-3		6-3	TERM
A-1	EMT										60'			2	THHN											66'						
A-2			20		2		19		2								20				19											
A-3			18		2		12		2								18			12												
				12	2		26		2							12				26												
				12	2	16			2							12					16											
				14	2	20			2							14																
		32			2	22			2									36		20												
		26			2			21	2								30			22												
				12	2			14	2							12					21											
				12	2			14	2							12				14												
				12	2			18	2							12				14												
			20		2		19		2									20		18												
			18		2		17		2								18					20										
A-4	EMT										48'			2	THHN						18											
A-5										29'	24'			4					29							52						
A-6		36			2		17		2									36					24									
		30			2		32		2								30				17											
				14	2		20		2							14				32												
				14	2		20		2							14				20												
				18	2			16	2											20												
		22			2			16	2								22					16										
				12	2			18	2							12					16											
				12	2			20	2							12					18											
				10	2		24		2							10					20											
		40			2		30		2									40			24											
			20		2		28		2								20			30												
A-7	EMT										56'			2	THHN	20				28						62						
A-8			20		2		24		2		24'			2		20							24									
A-9		30			2		18		2								30					18										
				14	2		18		2							14					18											
				18	2		22		2							18				22												
			18		2	28			2								18				28											
			20		2	20			2							20				20												
			22		2	10			2							22				10												
		40			2			20	2									40		20												
				12	2			14	2							12				14												
				12	2			14	2							12				14												
			24		2			22	2								24			22												
TOTAL		256	200	210	68	116	346	207	68	29'	219'			12	TOTAL	274	230	172	29	358	235	54	48			180						

Figure 6.8 Branch Circuit Takeoff Schedule.

Form E-10: Outlet Detail and Takeoff Schedule. Provisions are made on Form E-10 for counting the outlets according to type and location, and also for designating the types of support material associated with the outlet box. Provisions are also made on this form for recording receptacle, switch, telephone, junction, and fixture outlets. Designation is provided for wall, ceiling, or floor outlets. The type of construction in which the outlets occur is also designated on this form.

The type of material associated with the different types of outlet boxes is designated by a heavily outlined square and the number involved in the outlet count is inserted in the square. Each column is summarized to provide a total number for each type of material, which is posted at the foot of the column. All the totals of quantities are transferred from Form E-10 to Form E-19, under Branch Circuit Work Schedule, the only exception is the devices, plates, and covers, which are transferred to Form E-18, Finishing Tabulating Form.

Form E-10, filled out for a typical takeoff, is shown in Fig. 6.9.

Form E-11: Power System Takeoff Schedule. Form E-11 is used to record takeoff information for materials required for the wiring of motor outlets, power receptacle outlets, and motor control devices and equipment. A great deal of the work included in this category may be defined as power system branch circuit work.

Each circuit is identified on the form by the circuit number or the source of supply. The main categories shown on the form are the following:

- Motor services
- Power outlets
- Conductor
- Raceway and fittings

The listings for all conduit runs, equipment, and boxes are associated with the type of construction in which they are to be installed, and designation must be made for height of installation above 14 ft. The material requirements for outlets must be developed on Form E-10 as is done in the case of all other branch circuit–type installations. The quantity information on this form is summarized and transferred to the appropriate tabulating form, as follows:

- Feeder Conduit Tabulating Form
- Branch Circuit Tabulating Form
- Conductor Tabulating Form

Form E-10

OUTLET Detail and Takeoff Schedule

SHEET NO 1 OF 1
FLOOR NO 1 BLDG A
ESTIMATOR R.C.P.
DATE 11-8-84

JOB NAME Littleton Arts Center

TYPE		DESCRIPTION	QUANTITY	4" SQ BOX-1½	4" OCT BOX-1½	4 11/16 SQ BOX-2⅛	4" SQ BOX-2⅛	4" OCT BOX-2⅛	STEEL FLR BOX	ANCH PLAS	ASSY BOX FORM	BOLT TOG	BUSHING, GRC ½"	BUSHING, GRC ¾"	BUSHING, GRC 1"	CONN. WIRE	CONN. EMT 1/2"	CONN. EMT 3/4"	CONN. EMT 1"	COVER 4" SQ BLANK	COVER 4" SQ IND 1 G	COVER 4" SQ IND 2 G	COV PLAS 1 DEV	COV 3" OCT	COV WPF SW	HANG BOX J BAR	HANG BOX ADJ.	HANG BOX DEEP	LOCKNUT 1/2"	LOCKNUT 3/4"	LOCKNUT 1"	PLATE DEV 1 GANG BLANK	PLATE DEV 1 GANG D.R.	PLATE DEV 1 GANG SW	PLATE DEV 1 GANG TEL	RECEP DUP	RECEP SING	SWTCH 1P	SWTCH SW.	SCR 1"-8	STUD FIX	STUD MISC
OUTLET-FIXTURE	CEILING	Exposed on concrete																																								
OUTLET-FIXTURE	CEILING	Exposed on furred																																								
OUTLET-FIXTURE	CEILING	Furred flush	40+62+38 10+27		177											354	354							177				177														
OUTLET-FIXTURE	CEILING	Recessed in furred	卌 卌 卌 ||||	19												38	38	19		19																						19
OUTLET-FIXTURE	CEILING	Wood frame flush																																								
OUTLET-FIXTURE	WALL	Masonry flush	10+14+ 6+12+4	46												92	92						46																		46	
OUTLET-FIXTURE	WALL	Metal stud flush	70+21 46+10	147												294	294						147			147														294	147	
OUTLET-FIXTURE	WALL	Rein. concrete flush																																								
RECEPTACLE	DUPLEX	Exposed on concrete	卌 |||	8						8							16				8															8				8		
RECEPTACLE	DUPLEX	Masonry flush	卌 卌 ||| 卌 卌	23													46						23										23			23						
RECEPTACLE	DUPLEX	Metal stud flush	65+35+80 70+10+20				280												560				280					280					280			280						
RECEPTACLE	DUPLEX	Wood frame flush																																								
RECEPTACLE	SINGLE WALL FLR	Deck flush																																								
RECEPTACLE	SINGLE WALL FLR	Rein. concrete flush																																								
RECEPTACLE	SINGLE WALL FLR	Wood frame flush																																								
SWITCH	TUMBLER-WALL	Exposed on concrete	卌 |	6						6							12				6																	6		6		
SWITCH	TUMBLER-WALL	Masonry flush	卌 卌 ||	12													24						12					12						12				12		24		12
SWITCH	TUMBLER-WALL	Metal stud flush	20+15 10+43				88												176				88					88						88				88				
SWITCH	TUMBLER-WALL	Wood frame flush																																								
SWITCH	TUMBLER-WALL	Rein. concrete flush																																								
JUNCTION	WALL	Metal stud	4+1+4	9												18	18						9					9				9										9
JUNCTION	WALL	Reinforced concrete																																								
JUNCTION	WALL	Wood frame																																								
TELEPHONE	WALL	Masonry flush																																								
TELEPHONE	WALL	Wood frame flush	61+28+17 45+16	167													334						167			167									167							167
TELEPHONE	FLOOR	Deck flush	20+17+7						44				88																176						44							
TOTALS				437	177		368		44	14			88			796	1,228	19	736	19	14		772	177		314		506	176			9	303	100	211	311		106		332	193	167 19 21

Figure 6.9 Outlet Detail and Takeoff Schedule.

The quantities of items of equipment such as motor control devices are summarized and transferred directly to Form E-19, Pricing Sheet, under Equipment Work Schedule.

Typical information is shown on a copy of Form E-11 in Fig. 6.10.

Form E-12: Telephone System Takeoff Schedule. Form E-12 is used to record quantities of material for the telephone system, including panelboards, raceway, outlets, and all other items pertaining to the telephone system. Most telephone system hookups are installed by the utility, which means that the contractor provides all the rough-in materials, and the utility pulls the cables and installs the equipment, using the contractors No. 8 fish wire. Type of construction and height of installation must be recognized by the estimator.

Quantities are summarized and the totals are transferred to tabulating forms as follows:

- Panelboard quantities are entered directly on Form E-19.
- Feeder conduit is entered on Form E-15.
- Branch circuit conduit is entered on Form E-16.
- Wire quantities (if any) are entered on Form E-17.

The estimator must use Form E-10 to develop the information on mounting material required for the outlets.

Form E-12 is shown in Fig. 6.11, filled in with typical information.

Form E-13: Special Systems Takeoff Schedule. Form E-13 is used to record takeoff quantities for equipment, control, raceway, wire, and outlets for special systems such as alarm systems, annunciator systems, sound systems, and so forth. The estimator must use Form E-10 to develop the information on materials required to support the outlets indicated in these systems. Types of construction and height of installation must be recognized.

The total quantities of material developed on this form must be transferred to tabulating forms as follows:

- Equipment items are entered directly on Form E-19.
- Raceway and fittings are entered on Form E-16.
- Wire or cable is entered on Form E-17.
- Outlet material except for devices and plates is entered on Form E-16.
- Devices and plates are entered on Form E-18.

Form E-13 is shown in Fig. 6.12, filled in with typical items and quantities.

Form E-11

Power System Takeoff Schedule

JOB NAME Littleton Arts Center

SHEET NO. 1 OF 1
FLOOR NO. 1 BLDG. A
ESTIMATOR R.C.P.
DATE 11-8-84

SIGNALS USED ON FORMS FOR FLAGGING DEVIATIONS FROM STANDARDS

HEIGHT OF INSTALLATION		CONDUIT PARALLEL RUNS		WIRE PULLS IN PARALLEL	
Height	Designation	Number	Designation	Number	Designation
15-25 ft	●	2	2r	2	2p
26-35-ft	✕	3	3r	3	3p
36-50 ft	✓	4	4r	4	4p

MISCELLANEOUS SIGNALS
Density of Items — d
Number of Men — n
Note: Designation should be made in upper or lower left-hand corner of quantity column square.

IDENTIFICATION			MOTOR SERVICE							POWER OUTLETS				CONDUCTOR					RACEWAY		LENGTH RUN									
SOURCE SUPPLY	CT. NO.	LOCATION	HP	PH.	VOLTS	SAFETY SWITCH	STARTER	P.B. STA.	FLEX CONN.	AMPS	VOLTS	POLES	FL. OR SURF.	SIZE	TYPE	NO. OF COND.	LENGTH OF PULL	TERM	SIZE	TYPE	EMB'D	EXP.	FURRED	FRAME	HANGER		ELLS	FITT'G	BOXES	TERM
PPA	1	Pump	20	3	208		Magnetic	S/S	1"					#6	THW	3	65'	6	1"	GRS					60'		2	1-LB		4
PPA	2	Comp.	50	3	208		Magnetic	S/S	2"					4/0	THW	3	85'	6	2"	GRS	80'						2			2
PPA	3	Fan	1	3	208		Magnetic	S/S	1/2"					12	THW	3	48'	6	1/2"	EMT		48'						1-LB		2
PPA	4	Plugs								20	208	3	S	12	THW	3	60'	6	1/2"	EMT		60'								2
										10	208	2	S	12	THW	2	12'	4	1/2"	EMT		12'								2
PPB	1	Roof	5	3	208	RT	R.T. Magnetic	S/S	1/2"					12	THW	3	55'	6	1/2"	EMT			55'					1-EMT	4 sq WPF	4
PPB	2	Plugs								10	208	2	S	12	THW	2	44'	4	1/2"	EMT			44'					1-LL		2
										10	208	2	S	12	THW	2	10'	4	1/2"	EMT			10'							2
										10	208	2	S	12	THW	2	10'	4	1/2"	EMT			10'							2
PPB	3	El. Ht	40kW	3	208		135A Cont	TH	1 1/2					1/0	THW	3	76	6	1 1/2	GRS					80'		2			2
PPB	4	El. Rge	36kW	3	208		3 Boost Trans		1 1/2					1/0	THW	3	102'	12	1 1/2	GRS	102'						3			2
														6	THW	2	20	12	3/4	EMT		45							3-4" sq	6

Figure 6.10 Power System Takeoff Schedule.

Form E-12

Telephone System Takeoff Schedule

JOB NAME Littleton Arts Center

SHEET NO. 1 OF 1
FLOOR NO. 1 BLDG. A
ESTIMATOR R.C.P.
DATE 11-9-84

SIGNALS USED ON FORMS FOR FLAGGING DEVIATIONS FROM STANDARDS

HEIGHT OF INSTALLATION		CONDUIT PARALLEL RUNS		WIRE PULLS IN PARALLEL	
Height	Designation	Number	Designation	Number	Designation
15-25 ft	•	2	2r	2	2p
26-35-ft	x	3	3r	3	3p
36-50 ft	✓	4	4r	4	4p

MISCELLANEOUS SIGNALS
Density of Items — d
Number of Men — n
Note: Designation should be made in upper or lower left-hand corner of quantity column square.

PANELBOARDS						RACEWAY		LENGTH OF RUN										OUTLETS						WIRE OR CABLE						MISC.	
NO.	SIZE	FL.	SURF.	TYPE CONST.	WOOD BACK	SIZE	TYPE	EMB'D	EXP.	FURR'D.	FRAME	HANGER	MAS'Y.	ELLS	FITT.	BOXES	TERM	TYPE	FL.	SURF	TYPE CONST.	FLOOR	UFD	SIZE	TYPE	NO	NO./PAIR	LENGTH OF PULL	TERM	TERM BLOCK	
1	36"x30"x4"		X	Masonry	3/4" Ply	2"	GRS	80'						2			2							#8	Iron						
2	30"x24"x4"		X	Wood	3/4" Ply	1 1/2"	GRS		60'			50'		2			2							#8	Iron						
						3/4"	EMT				30'						2	Wall	F		Frame										
						3/4"	EMT				24'						2	Wall	F		Frame										
						3/4"	EMT	18									2	Flr.	F		Deck										
						1/2"	EMT				28'						2	Wall	F		Frame										
						3/4"	EMT				36'						2	Wall	F		Frame										
						1/2"	EMT						12'				2	Wall		S	Masy										
						3/4"	EMT						90'				2	Wall		S	Masy'										

Figure 6.11 Telephone System Takeoff Schedule.

Form E-13

Special Systems Takeoff Schedule

JOB NAME Littleton Arts Center

SHEET NO 1 OF 1
FLOOR NO 1 BLDG A
ESTIMATOR R.C.P.
DATE 11-10-84

SIGNALS USED ON FORMS FOR FLAGGING DEVIATIONS FROM STANDARDS

HEIGHT OF INSTALLATION		CONDUIT PARALLEL RUNS		WIRE PULLS IN PARALLEL	
Height	Designation	Number	Designation	Number	Designation
15-25 ft	•	2	2r	2	2p
26-35 ft	×	3	3r	3	3p
36-50 ft	✓	4	4r	4	4p

MISCELLANEOUS SIGNALS
Density of Items — d
Number of Men — n

Note: Designation should be made in upper or lower left-hand corner of quantity column square

TYPE SYSTEM	EQUIPMENT DESC	SIZE	WGT.	MOUNT	WIRE TERM	RACEWAY SIZE	TYPE	LENGTH OF RUN EMB'D	EXPOSED	FRAME	HANGER	ELLS	FITT.	TERM.	PULL BOXES	WIRE OR CABLE SIZE	TYPE	LENGTH PULL	NO./WIRE	NO./PAIRS	TERM.	TERM. BLOCK	OUTLETS QUAN	TYPE	FL/SF	AMPS	VOLTS	POLES
																18	TF	65'	18-18		18	20 Pr.						
Alarm	Signal	10"x7"x4"	12	S	18-18	1"	GRS			65'				2	1-4"Sq	18	TF	60		18-3	6		1	4"Sq	S		12	3
	Panel															18	TF	80		18-3	6		1	4"Sq	S		12	3
	Flr. Mats	2'x4'	3	S		1/2"	EMT			70'						18	TF	27		18-3	6		1	4"Sq.	S		12	3
						1/2"	EMT			25'						18	TF	43		18-3	6		1	4"Sq	S		12	3
						1/2"	EMT			40'						18	TF	55		18-3	6		1	4"Sq	S		12	3
						1/2"	EMT			50'						18	TF	46		18-3	6		1	4"Sq.	S		12	3
						1/2"	EMT			45'																		
Alarm	Trans.	10VA		S	2-16	1/2"	EMT			20'						16	TF	22'		16-2	4							
Alarm	Relay	10"x8"x4"	10	S	15-18	3/4"	EMT			20'				2	1-4"Sq	18	TF	24'		18-3	6							
	Panel					1/2"	EMT			40'						18	TF	45		18-3	6							
						1/2"	EMT			45'						18	TF	50		18-3	6							
						1/2"	EMT			45'						18	TF	50		18-3	6							
						1/2"	EMT			38'						18	TF	42		18-3	6							

Figure 6.12 Special Systems Takeoff Schedule.

Principle 6: Tabulate the Takeoff by Similar Items

Many identical items of material appear on the takeoff forms, and it is essential to accumulate them in order to accomplish the following important functions in the estimating process:

- To accumulate identical items of material installed in the same type of construction and at the same height
- To provide a transition from system takeoff to construction work schedule
- To provide a summary of quantities of identical materials for listing on pricing sheets according to work schedule
- To eliminate duplicate listing and pricing
- To provide the vehicle for listing items of material on Bills of Material by work schedule for job management

Figure 6.13 illustrates the process of transferring information from the takeoff forms, which are organized by systems; to the tabulating forms, which are set up by work schedule; and thence to pricing sheets, which are also organized by work schedule. The figure shows that there are nine takeoff forms, five tabulating forms, and seven work schedules. It should be noted that there is no tabulating form for lighting systems and equipment because all the items in those categories are recorded on their respective takeoff form and do not require tabulating.

The following tabulating forms are discussed:

- Form E-14, Embedded Tabulating Form
- Form E-15, Feeder Conduit Tabulating Form
- Form E-16, Branch Circuit Tabulating Sheet
- Form E-17, Conductor Tabulating Sheet
- Form E-18, Finishing Work Tabulating Sheet

Form E-14: Embedded Tabulating Form. The quantities of embedded conduit and fittings are tabulated separately for the purpose of scheduling material to the job as a job management function. It is important for the Bill of Material to list these items separately for reasons of scheduling. Form E-14 is used to record quantities from the following takeoff forms:

- Form E-3, Service, Metering, and Grounding
- Form E-7, Feeder and Busway
- Form E-9, Branch Circuit

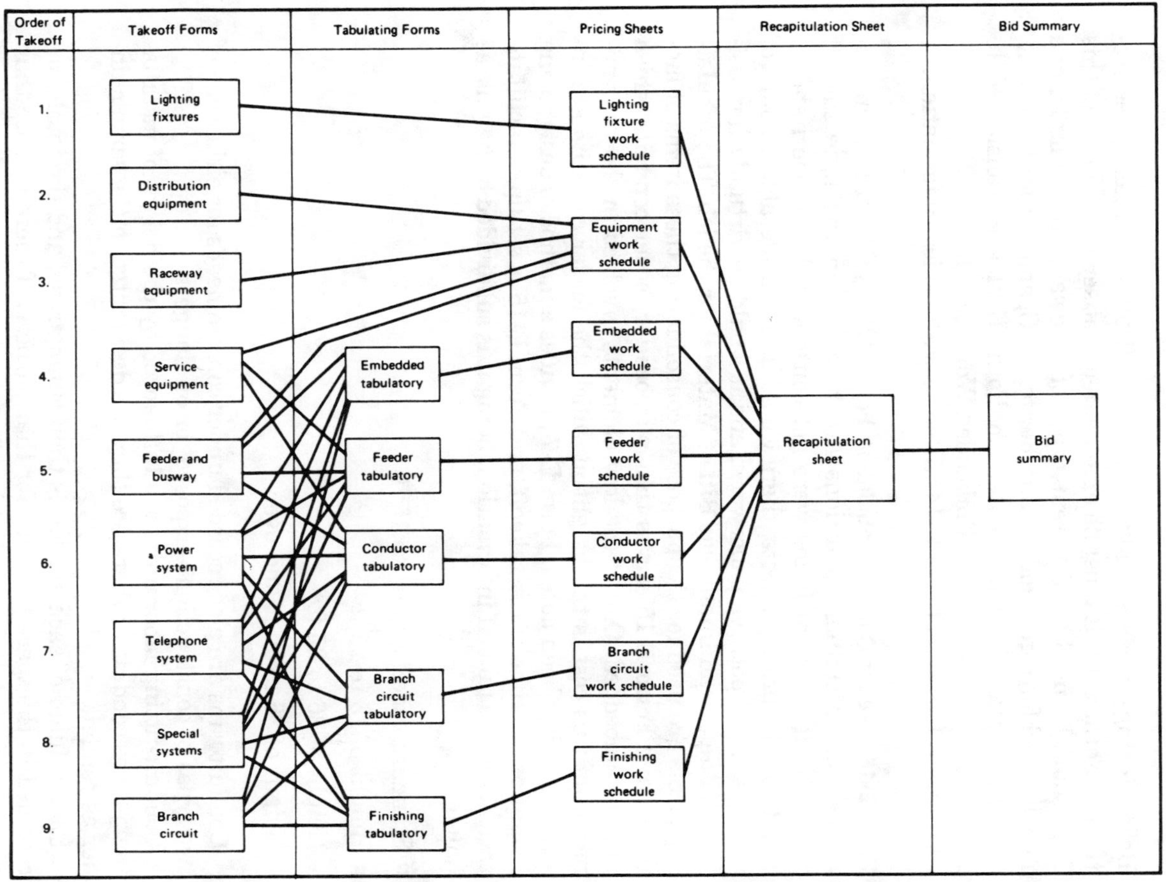

Figure 6.13 The relationship between items of material on takeoff forms, tabulating forms, and pricing sheets.

- Form E-11, Power System
- Form E-12, Telephone System
- Form E-13, Special Systems

Boxes, fittings, sleeves, and trenching are listed in the last three sections of Form E-14. It must be emphasized that separate sheets of this form must be used for different types of raceway. For example, one may be used for rigid conduit, another for PVC, another for fiber duct.

Items of material summarized on Form E-14 are transferred to Form E-19, Pricing Sheet—Embedded Work Schedule.

Figure 6.14 shows Form E-14, filled in with typical information.

Form E-15: Feeder Conduit Tabulating Form. Many identical items installed under similar conditions are recorded on the nine takeoff forms. If all identical items were not consolidated into one item for listing on the pricing sheets, there would be a considerable amount of unnecessary repetitive paperwork, and it would be virtually impossible to generate an intelligent Bill of Material for the job. The tabulating forms are used for the purpose of consolidating these items of material, and Form E-15 serves that purpose for feeder conduits which are not embedded. One of the key elements involved in the determination of this cost is the height at which the feeder conduits are installed. As has been noted, Form E-7 provides a method of designating height at the time of the takeoff, and Form E-15 is set up to facilitate transfer of the takeoff information, being it is subdivided by height, as follows:

- Standard conditions up to 15 ft
- Height of 15 to 25 ft
- Height of 26 to 35 ft
- Height of 36 to 50 ft and over

Conduit runs are also recorded under column heads for either one of two types of construction—exposed, or on hangers.

Conduit-fitting labor units are influenced by the height of installation just as conduit is; therefore they are designated in the same manner as conduit.

Each column of each section is then summarized and the totals are recorded in the space provided for that purpose. If there is more than one type of conduit involved, it is necessary to use another form sheet for the second type.

Each of the totals for each condition of installation is entered separately on Form E-19, Pricing Sheet—Feeder Conduit Work Schedule.

Form E-14.

Embedded Tabulating Form

JOB NAME *Littleton Arts Center*

SHEET NO. *1* OF *1*
FLOOR NO. *1* BLDG. *A*
ESTIMATOR *R.C.P.*
DATE *11-11-84*

NOTE: Use separate sheets of this form for each different type of conduit.

SYSTEM REFER'NCE	RACEWAY — TYPE: *GRS*																														BOXES		FITTINGS		SLEEVES TRENCH			
	1/2"			3/4"			1"			1¼"			1½"			2"			2½"			3"			3½"													
	QUAN.	BENDS	TERM	QUAN.	BENDS	TERM	QUAN.	ELLS	TERM	QUAN.	ELLS	TERM	QUAN.	ELLS	TERM	QUAN.	ELLS	TERM	QUAN.	ELLS	TERM	QUAN.	ELLS	TERM	QUAN.	ELLS	TERM	QUAN.	ELLS	TERM	QUAN.	TYPE	QUAN.	SIZE	QUAN.	SIZE	QUAN.	SIZE
Service																									*2@30'* *2@30'*	*2* *2*	*2* *2*				*1*	*C.T. Cab*						
Feeder																			*120'*	*2*	*2*												*1*	*Exp. 2"*	*1*	*2"*		
Br. Ct.	*256'*	*6*	*12*																																			
Power													*110'*	*2*	*2*	*80'*	*2*	*4*															*1*	*Exp. 1½*	*1*	*1½"*		
Tel.																*80'*	*2*	*2*																			*80'*	*12"x 24"*
Special System							*70'*	*2*	*4*																													
TOTAL							*70'*	*2*	*4*				*110'*	*2*	*2*	*160'*	*4*	*6*	*120'*	*2*	*2*				*2@60'*	*4*	*4*				*1*	*CT Cab*	*1-* *1-*	*Exp 1½* *2*	*1* *1*	*1½"* *2"*		

Figure 6.14 Embedded Tabulating Form.

Figure 6.15 shows Form E-15, Feeder Conduit Tabulating Form, with typical information recorded.

Form E-16: Branch Circuit Tabulating Sheet. Quantities of identical branch circuit material items appear on several takeoff forms, and it is necessary to accumulate totals of these on Form E-16. Form E-16 is divided into eight sections: raceway, cable, outlet boxes, plaster covers, box hangers, EMT fittings, GRS fittings, and cable fittings. There is space on the form for three sizes of raceway, and under each size of conduit there are columns for three types of construction: exposed, frame, or furred.

Each form sheet accommodates only one type of conduit. If there are others it is necessary to use a separate form sheet.

Figure 6.16 shows Form E-16, filled in with typical information.

Form E-17: Conductor Tabulating Sheet. Form E-17 is used to record quantities of wire and cable from takeoff forms. The form is divided into five sections, as follows:

- The first section is used to record quantities of branch circuit wire according to sizes (12, 10, and 8), and also according to the number of conductors in the pull, which is two, three, four, or five. Wire connectors are recorded in the end column.
- Four sections of the form are used to accumulate totals of feeder cable according to length of pull and number of conductors, with lengths of pull as follows: pulls not over 50 ft; pulls not over 100 ft; pulls not over 150 ft; pulls exceeding 150 ft.

Columns are provided for 3- and 4-wire runs.

The individual length of pull is recorded separately in the proper section. Each column is then summarized according to length, and is converted to total feet of wire by multiplying the length of pull by the number of wires in the pull. Connectors are summarized in the end column of each section.

When different types of wire are involved, it is necessary to use a separate form for each type. Figure 6.17 shows Form E-17, with typical information recorded.

Form E-18: Finishing Work Tabulating Sheet. Form E-18 is used to record the quantities of material used in the finishing stages of electrical construction work. There are 10 sections of this form, devoted to device plates, receptacles, switches, cord caps, cord types, pushbutton controls, control transformers, thermostats, underfloor duct risers, and thermal overload switches.

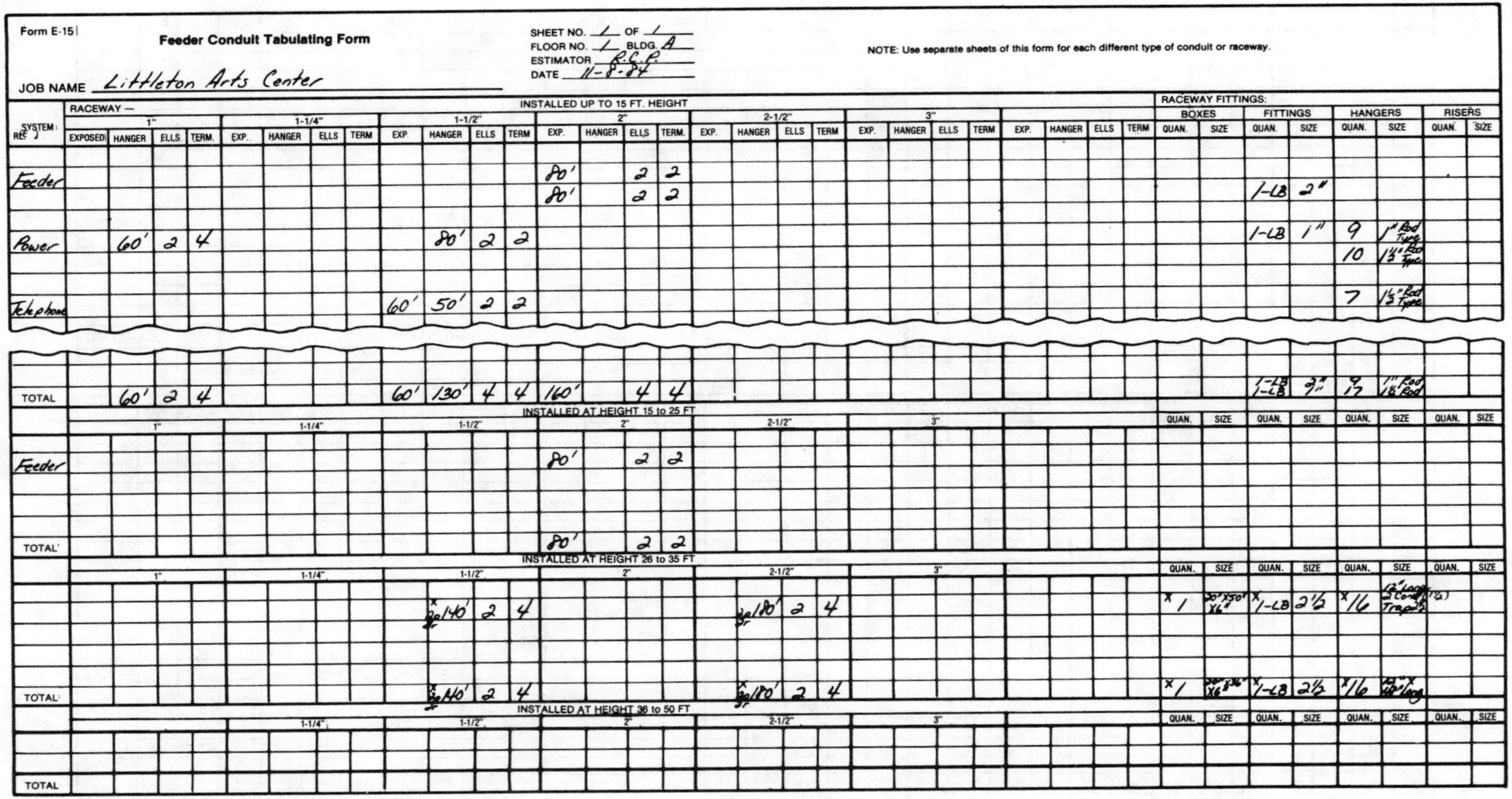

Form E-15 — **Feeder Conduit Tabulating Form**

JOB NAME: Littleton Arts Center

SHEET NO. 1 OF 1
FLOOR NO. 1 BLDG. A
ESTIMATOR R.C.P.
DATE 11-8-84

NOTE: Use separate sheets of this form for each different type of conduit or raceway.

RACEWAY — INSTALLED UP TO 15 FT. HEIGHT

SYSTEM REF.	1" HANGER	1" ELLS	1" TERM.	1-1/2" EXP.	1-1/2" HANGER	1-1/2" ELLS	1-1/2" TERM	2" EXP.	2" ELLS	2" TERM.	FITTINGS QUAN.	FITTINGS SIZE	HANGERS QUAN.	HANGERS SIZE
Feeder								80'	2	2				
								80'	2	2	1-LB	2"		
Power	60'	2	4		80'	2	2				1-LB	1"	9	1" Rod Type
													10	1½" Rod Type
Telephone				60'	50'	2	2						7	1½" Rod Type
TOTAL	60'	2	4	60'	130'	4	4	160'	4	4	1-LB 1-LB	2" 1"	9 17	1" Rod 1½" Rod

INSTALLED AT HEIGHT 15 to 25 FT

SYSTEM REF.	2" EXP.	2" ELLS	2" TERM.
Feeder	80'	2	2
TOTAL	80'	2	2

INSTALLED AT HEIGHT 26 to 35 FT

SYSTEM REF.	1-1/2" HANGER	1-1/2" ELLS	1-1/2" TERM	2-1/2" HANGER	2-1/2" ELLS	2-1/2" TERM	BOXES QUAN.	BOXES SIZE	FITTINGS QUAN.	FITTINGS SIZE	HANGERS QUAN.	HANGERS SIZE
	× [illegible] 140'	2	4	× [illegible] 180'	2	4	×1	30"×50"×6"	×1-LB	2½	×16	12" Long 3 Cond (1½) Trapeze
TOTAL	× [illegible] 140'	2	4	× [illegible] 180'	2	4	×1	30"×36"×6"	×1-LB	2½	×16	[illegible] × 18" long

INSTALLED AT HEIGHT 36 to 50 FT

SYSTEM REF.	
TOTAL	

Figure 6.15 Feeder Conduit Tabulating Form.

Form E-16

Branch Circuit Tabulating Sheet

JOB NAME Littleton Arts Center

SHEET NO. 1 OF 1
FLOOR NO. 1 BLDG. A
ESTIMATOR RCP
DATE 11/9/84

SYSTEM	RACEWAY — TYPE: E.M.T. 1/2" EXPOSED	1/2" FRAME	1/2" FURRED	3/4" EXPOSED	3/4" FRAME	3/4" FURRED	1" EXPOSED	1" FRAME	1" FURRED	ELLS	CABLE/GROUND - TYPE: NUMBER OF CONDUCTORS 12-2	12-3			6-3	TERM	BOXES, OUTLET 4" SQUARE 1½" DP	4" SQUARE 2⅛" DP	4" SQ BRACKET	4 11/16" SQ 1½" DP	4 11/16" SQ 2⅛" DP	4" OCT 1½" DP	4" OCT 2⅛" DP	SEC SW.	CONC RING	SING FL.BX	COVERS, PLASTER 1 GANG 1/2"	1 GANG 3/4"	2 GANG 1/2"	2 GANG 3/4"	OCT. COV.	CONC. COV.
BR CT	510	13455			285'												341	42	201	16		150					510	40		42	140	10
TEL	120'	120'		50'													6									1	6					
SPEC SYSTEM					20'																											
POWER SYSTEM	130'	120'		50'													5										5					
TOTALS	760	13,695	1504	100	305	44											352	42	201	16		150				1	521	40		42	140	10

SYSTEM HANGERS BOX	EXP'N 18"	ST. BAR 18"	DP. BAR 18"	SH. BAR 18"	JACK BAR	DP. BAR 26"	CADDY	FITTINGS, EMT CONNECTORS 1/2"	CONNECTORS 3/4"	CONNECTORS 1"	COUPLINGS 1/2"	COUPLINGS 3/4"	COUPLINGS 1"	CLAMPS, 1 HOLE 1/2"	CLAMPS, 1 HOLE 3/4"	CLAMPS, 1 HOLE 1"	STRAPS, 2 HOLE 1/2"	STRAPS, 2 HOLE 3/4"	STRAPS, 2 HOLE 1"	FITTINGS, G.R.S. LOCKNUTS 1/2"	LOCKNUTS 3/4"	LOCKNUTS 1"	BUSHINGS 1/2"	BUSHINGS 3/4"	BUSHINGS 1"	2 HOLE STRAPS 1/2"	2 HOLE STRAPS 3/4"	2 HOLE STRAPS 1"	FITTINGS, CABLE STRAP NAILS	BOX CONN.	NAILS	STRAPS
BR CT	106		150				493	1470	30		1400	38		51																		
TEl	6							16	6		25	5		12	5																	
SPEC System												2																				
Power System								18	6		25	5		13	5																	
TOTALS:	112		150				493	1504	42		1450	50		76	10																	

Figure 6.16 Branch Circuit Tabulating Sheet.

Form E-17

Conductor Tabulating Sheet

SHEET NO. 1 OF 1
FLOOR NO. 1 BLDG. A
ESTIMATOR R.C.P.
DATE 11-9-84

NOTE: Use separate sheets of this form for each different type of conductor.

JOB NAME Littleton Arts Center

SYSTEM	TYPE WIRE	#12 - LENGTH PULL					#10 - LENGTH PULL				#8 — LENGTH PULL				CONNECTORS	
		2 COND.	3 COND.	4 COND.	5 COND.	TERM	2	3	4	TERM	2	3	4	TERM	SIZE	QUAN
Br Ct Sh 1	THW	2416	762	582	51	310										
Br Ct Sh 2	THW	1910	615	409	40	265										
Br Ct Sh 3	THW	2445	809	612	58	350										
Br Ct Sh 4	THW	1642	595	410	39	195										
Pwr 1	THW	745	703	381	111	514										
TAKEOFF FOOTAGE		9178	3484	2394	299											
LINEAL FOOTAGE		18356	10452	9576	1495	1634										

SYSTEM	TYPE WIRE	#12 LENGTH PULL					#10 LENGTH PULL				#8 LENGTH PULL				CONNECTORS	
		2 COND	3 COND.	4 COND	5 COND	TERM	2	3	4	TERM	2	3	4	TERM	SIZE	QUAN.

FEEDER CABLE: TYPE THW — LENGTH OF PULL — NOT OVER 50 FT.

SYSTEM	#6		#4		#2		#1		#1/0		#2/0		#3/0		#4/0		#250 MCM		#300 MCM		#350 MCM		#400 MCM		#500 MCM		#8	CONNECTORS-LUG		
	3 COND.	4 COND	3 COND	4 COND	3	4	3	4	3	4	3	4	3	4	3	4	3	4	3	4	3	4	3	4	3	4	2 COND.	TYPE	SIZE	QUAN.
Feeder	65																										20	Comp	6	10
SUM OF PULLS	65																										20			
TOTAL FOOTAGE	195																										40			

FEEDER CABLE: TYPE THW — LENGTH OF PULL — NOT OVER 100 FT.

SYSTEM	3	4	3	4	3	4	3	4	3	4	3	4	3	4	3	4	3	4	3	4	3	4	3	4	3	4		CONNECTORS-LUG		
Feeder									82'			86'			65													Comp	1/0	12
									108'						80													Comp	2/0	8
															85													Comp	4/0	18
SUM OF PULLS									190			86			230															
TOTAL FOOTAGE									570			344			690															

FEEDER CABLE: TYPE THW — LENGTH OF PULL — NOT OVER 150 FT.

SYSTEM	3	4	3	4	3	4	3	4	3	4	3	4	3	4	3	4	3	4	3	4	3	4	3	4	3	4		CONNECTORS-LUG		
Feeder									2-146									136										Comp	1/0	6
																												Comp	250	8
SUM/PULLS									2-146									136												
TOTAL FOOTAGE									2-438									544												

FEEDER CABLE TYPE THW — LENGTH OF PULL — EXCEEDING 150 FT.

SYSTEM	3	4	3	4	3	4	3	4	3	4	3	4	3	4	3	4	3	4	3	4	3	4	3	4	3	4		CONNECTORS-LUG		
Feeder																		2-181										Comp	250	6
SUM/PULLS																		181												
TOTAL																		2-724												

Figure 6.17 Conductor Tabulating Sheet.

All the items on Form E-18 are summarized and entered on Form E-19, Pricing Sheet—Finishing Work Schedule.

Figure 6.18 shows Form E-18, filled in with typical information.

Tabulating by work schedule in the manner advocated in this section is no more laborious than it would be as a batch operation. The advantage is that the work schedules provide a basis for scheduling the

Form E-18

Finishing Work Tabulating Sheet

SHEET NO. 1 OF 1
FLOOR NO. 1 BLDG. A
ESTIMATOR RCP
DATE 11-10-84

JOB NAME Littleton Arts Center

SYSTEM	PLATES — FINISH: IVORY 1G. DUP. RECEP.	2G. DUP. RECEP.	2G. COMB SW. REC.	1G. SING RECEP.			1G. TUMB. SWITCH	2G. TUMB. SWITCH	3G. TUMB. SWITCH		1G. TEL.	1 GANG BLANK	
BrCt	84						48	12	4				
Power				6									
Tel											6		
TOTAL	84			6			48	12	4		6		

SYSTEM	RECEPTACLES — TYPE: DUPLEX	SINGLE	1φ 3 WIRE 10A.	1φ 3 WIRE 20A.	3φ 4 WIRE 20A.	3φ 4 WIRE 30A.	3φ 4 WIRE 50A.	FINISH: CLOCK	PLAS. PROT.	SWITCHES — TYPE: SPEC. FIN: IVORY 1 POLE	2 POLE	3 WAY	4 WAY		PHOTO CELL
BrCt	84								168	54		8	2		
Power			4		1			4							
TOTAL	84		4		1			4	168	54		8	2		

SYSTEM	CORD CAPS 10A. 3 W.	20A. 3W	20A 4W	30A 3W	30A 4W	50A 3W	50A 4W	CORD — TYPES: 18-2	18-3	18-4	16-2	16-3	16-4	
BrCt														
Power	4		1											
TOTAL	4		1											

SYSTEM	PUSH BUT. CONTROLS QUAN.	SIZE	TYPE	CONTROL TRANSFORMER QUAN.	SIZE	TYPE	THERMOSTAT QUAN.	SIZE	TYPE	U.F.D. RISERS QUAN.	SIZE	TYPE	TO SWITCHES QUAN.	SIZE	TYPE
BrCt				1	50VA	M	4	REG	GP						
Power	4	5/5	GP												
TOTAL	4	5/5	GP	1	50VA	M	4	Reg	GP						

Figure 6.18 Finishing Work Tabulating Sheet.

material to the job in an orderly manner; the result is that material need not be placed on the job until it is needed. The tabulating process is also necessary in order to convert the material listing from systems to work schedules.

Principle 7: Price and Factor the Estimate

Pricing is the process of listing on Form E-19, Pricing Sheet, according to work schedule, all identical items from the tabulating forms so that the material can be priced and labor units assigned accordingly. (See Fig. 6.19.) The work schedule on which items are listed is determined by the stage of construction at which they are to be installed, as follows:

Footings and foundations	Embedded Work Schedule
Superstructure	Feeder System Work Schedule
Floors	Equipment Work Schedule
Interior walls and ceilings	Branch Circuit Work Schedule
Building enclosure	Conductor and Wire Work Schedule
Trimming	Lighting System Work Schedule
Painting and decorating	Finishing Work Schedule

A separate copy of the Pricing Sheet is used for listing each work schedule, which means that there will be seven copies of Form E-19, Pricing Sheet, for each estimate. Each copy should be clearly marked so that it is clear which work schedule it refers to. The material listed on the Pricing Sheet is transferred directly from a takeoff form or from a tabulating form. All items are listed so that pricing, laboring, and factoring may be properly done, and each sheet is totaled separately. It is desirable to list material items first by name and second by size and type. It is also necessary to separate identical items by type of construction, number of conductors, and runs and pulls in parallel. For example:

- Footages of No. 12 wire must be listed according to the number of wires in the run (e.g., 2, 3, 4, or 5).
- Feeder wire listings must recognize the length of the pull, number of conductors, and parallel pulls.
- Conduit listings must recognize height factor, type of construction, and parallel runs.

In all cases, only standard labor units in man-hours must be entered in the proper column under labor cost; deviations from standards are handled by applying a factor to the standard unit. In addition to pro-

Form E-19

Pricing Sheet

JOB NAME Littleton Arts Center

SHEET NO. 1 OF 1
WORK SCHEDULE: Conductor
ESTIMATOR R.C.F.
DATE 11-12-84

STANDARD LABOR UNIT BASED ON CONDITIONS OF INSTALLATION AND USAGE	LENGTH/PULL	MULTIPLIER	PULLS IN PARALLEL	MULTIPLIER	HEIGHT FACTOR	MULTIPLIER	CONDUIT PARALLEL RUNS	MULTIPLIER	CONC. FACTOR	NUMBER OF MEN FACTOR
	50 ft	1.50	1	1.00	Scaffold 15'-25'	1.50	2	0.85	0.70 TO 0.87	1.50 TO 2.30
	100 ft	1.00	2	.90	Scaffold 26'-35'	1.70	3	0.80		
	150 ft	.80	3	.85	Scaffold 35'-50'	2.00	4	0.77		
	200 ft	.70	4	.80	Ladder 20'-22'	1.70	5	0.74		
	300 ft	.60	5	.78	Scissor 15'-30'	1.30				

REF NO.	DESCRIPTION	QUANTITY	MATERIAL COST: UNITS	PER	EXTENSION	LABOR UNIT MAN-HRS.	CONDUCTOR FACTORS: NO. OF COND'R	LENGTH OF PULL	L/PULL FACTOR	NUMBER PARALLEL PULLS	PULLS MULTIPLIER	CONDUIT FACTOR: INSTAL'D HEIGHT	HEIGHT FACTOR MULTIPLIER	1.30 EMBED'D	1.10 EXPOSED	1.10 FURRED HANGER TRENCH	1.45 WOOD FRAME	COND PAR'L RUN: NUMBER IN PAR'L	COND RUNS MULTIPLIER	CONCENTRA'N: QUAN.	CONC. FACTOR	NO/MEN FACTOR: EXCESS WORK	NUMBER OF MEN FACTOR	FACTOR'D LABOR UNIT	LABOR MAN HOURS
E-17	Wire, #12 THW. Copper	4,356'	33.60	M	616.76	.0086	2																	.0086	157.76
E-17	Wire, #12 THW. Copper	9,426'	33.60	M	316.71	.0065	3																	.0065	61.27
E-17	Wire, #12 THW. Copper	9,576'	33.60	M	321.75	.0060	4																	.0060	57.46
E-17	Wire, #12 THW. Copper	1,495'	33.60	M	50.23	.0055	5																	.0055	8.22
E-17	Wire, #6 THW. Copper	40'	137.00	M	5.48	.0167	2	20'	1.50															.025	1.00
E-17	Wire, #6 THW. Copper	195'	137.00	M	26.71	.0125	3	65'	1.50															.0187	3.65
E-17	Wire, #1/0 THW. Copper (2 pulls)	570'	455.00	M	259.35	.0244	3	93'	1.00	2	0.90													.0220	12.54
E-17	Wire, #1/0 THW. Copper (2 pulls)	438'	455.00	M	199.29	.0244	3	146'	0.85															.0207	9.07
E-17	Wire, #2/0 THW. Copper	344'	560.00	M	192.64	.0250	4	86'	1.00															.0250	8.60
E-17	Wire, #4/0 THW. Copper (3 pulls)	690'	870.00	M	600.30	.0395	3	77'	1.00	3	0.85													.0335	23.12
E-17	Wire, #250 MCM THW. Copper	544'	1,065.00	M	579.36	.0382	4	136'	0.85															.0325	17.68
E-17	Wire, #250 MCM THW. Copper (2 pulls)	724'	1,065.00	M	771.06	.0382	4	181'	0.70	2	0.90													.024	17.37
E-17	Conn, Scotchlok 16/10 Copper	1634	3.15	C	51.47	.03 EA																			49.02
E-17	Conn, Scotchlok 12/6 Copper	10	12.30	C	1.23	.04 EA																			0.40
E-17	Lug, Comp Type #1/0 Copper	18	87.13	C	15.68	.23 EA																			4.14
E-17	Lug, Comp Type #2/0 Copper	8	102.88	C	8.30	.25 EA																			2.00
E-17	Lug, Comp Type #4/0 Copper	18	138.50	C	24.93	.30 EA																			5.40
E-17	Lug, Comp Type 250 Copper	14	161.13	C	22.56	.37 EA																			5.18
	TOTAL MATERIAL COST (THIS SHEET)				$4,063.71	(NOTE: USE A DIFFERENT PRICING SHEET FOR EACH OF SEVEN WORK SCHEDULES)																		TOTAL LABOR MANHOURS (THIS SHEET)	443.49

Figure 6.19 Pricing Sheet—Conductor Work Schedule.

viding space for entering quantity of material, material unit prices and extension, Form E-19 provides ample space for adjusting labor units and factoring for construction conditions which necessitate changes in the basic labor unit.

All pricing sheets are individually totaled for material cost and man-hours of labor, but totals should also be made for each work schedule. The totals for each work schedule are entered on Form E-20, Recapitulation Sheet, which is discussed in the next section.

Material pricing

It is an elementary fact of the contracting business that in order for the contractor to be competitive material must be purchased at the lowest possible price, and that the price must be firm for the duration of the job, which may be many months. Securing prices for material and equipment is a time-consuming task and the estimator should have substantial backup for this responsibility.

Pricing labor

Assigning proper labor hours to the estimate is far more complex than material pricing, but if the estimator properly flags elements of the takeoff and carries them through the tabulating process, they should be easily reconciled on the Pricing Sheet when the proper provisions are made on the Pricing Sheet for adjusting the basic labor units.

Seven main categories of labor unit adjustment are performed on Form E-19, Pricing Sheet; these are adjustments for:

- Length of pull for feeder conductors
- Parallel pulls for feeder conductors
- Parallel conduit runs
- Height of installation
- Type of construction
- Number of men
- Concentration

In each case, a multiplier is applied to the standard labor unit, which is posted in the first column of the labor section of the form.

Labor unit adjustments in detail

Length of pull. Standard labor units have been established for a 100-ft run for each size of conductor according to the number of wires in the pull. When the length of pull is more or less than 100 ft, a multiplier is applied to the SLU as indicated below.

Length of pull	Multiplier
150 ft	0.85
200 ft	0.70
300 ft	0.60
50 ft	1.50

Parallel pulls. When two different pulls are made from the same setup point, it is proper to apply a factor for parallel pulls even though the sizes of conductor are different or the number of wires is different. The multiplier which would be applied to the SLU in combination with the length-of-pull factor is shown below.

No. of parallel pulls	Multiplier
Two	0.90
Three	0.85
Four	0.80
Five	0.78

An example of the items which would be involved if there were three pulls in parallel of sizes 2/0 with 150-ft runs is shown below.

Wire, No. 2/0 THHN, four conductors	125 ft
Wire, No. 2/0 THHN, four conductors	145 ft
Wire, No. 2/0 THHN, four conductors	130 ft
Total	400 ft

Form E-19 provides for the use of a standard labor unit for 2 wire, 3 wire, 4 wire, etc., and for the application of a multiplier according to length of pull and/or parallel pulls. For example, if there were three parallel pulls, a factor of 0.85 would be shown in the proper column for each size affected. Thus, if the wire size was No. 2/0 for each 4-wire parallel pull, the figures appearing on the form would be as shown below.

Standard labor unit	0.024 man-hour per foot
Factor for 150-ft pulls	0.85
Three pulls in parallel	0.85
Total labor unit	0.0174 per foot
Total labor for 1800 ft	32 man-hours

If the standard labor unit had not been factored, the labor would have been 43 man-hours, and a 25.6 percent reduction in cost would not have been included in the estimate. It is not possible to be competitive if one fails to include such factoring.

Height of installation. Standard labor units provide for installation up to and including 14 ft. When this elevation is exceeded, it becomes necessary to apply a multiplier to compensate for additional time required to make the installation. The added time may manifest itself in reduced efficiency, or an additional helper for safety. The multipliers for various heights using various types of elevation devices are shown below.

Height of installation	Multiplier
Up to 15 ft	0.00
15 to 25 ft (scaffold)	1.50
26 to 35 ft (scaffold)	1.70
Over 35 ft (scaffold)	2.00
20 to 22 ft (ladder)	1.70
15 to 30 ft (scissor lift)	1.30

When applicable, these figures are posted in the second column under Conduit Factors and multiplied by the SLU to obtain the factored labor unit in the next to last column, which when multiplied by the quantity produces total labor hours.

Type of construction. The standard labor unit which is posted in column 1 of the labor section of Form E-19 is based on conduit being installed in free air with no bends or offset. When branch circuit conduit is installed in other types of construction, the cost of installation is increased by the difficulty of fitting it into a structure which requires bends and offsets. The multipliers for this factor are shown below.

Type of construction	Multiplier
Deck	1.30
Exposed	1.15
Furred ceiling	1.10
Metal frame	1.40
Wood frame	1.45
Conduit on hangers	1.10
Flush-in masonry	1.45
Trench	1.10

These multipliers are based on the use of the following number of bends and offsets:

Types of construction	Number of bends and offsets per 100 ft
Deck	Two bends and two offsets
Exposed	Four bends and four offsets
Furred ceiling	Two bends and two offsets
Metal frame	Three bends and three offsets
Wood frame	Four bends and four offsets
Conduit on hangers	Two bends
Flush-in masonry	Three bends and three offsets
Trench	Two bends

In addition to the increased cost of branch circuit work caused by bends and offsets, there is another element of cost known as the *access factor*. These factors are included in the type-of-construction factors, and are shown in Table 6.3.

Parallel conduit runs. Standard labor units for conduit are based on a single run of conduit, but when it is possible to install multiple runs from the same setup, a multiplier may be applied to the SLU for all sizes involved, as shown below.

Number of parallel runs	Multiplier
Two	0.85
Three	0.80
Four	0.77
Five	0.74

This factor is posted in the proper space in the column headed up for that purpose.

Number of men. The number of men used to make an installation has considerable bearing on the magnitude of the labor unit, so a standard has been established for each category of material to indicate which operations are either one-man or two-man installations (refer to Table 6.5).

Sometimes local labor agreements are restrictive and it is impossible for the journeymen to conform to these estimating standards. In such cases it is necessary for the estimator to apply factors to the labor units as required. When a standard labor unit which is based on a one-man operation is converted for a two-man installation, the following time-consuming elements of the labor unit are doubled:

- Moving to the point of usage—6 percent on average
- Tooling up—4 percent on average
- Measuring and layout—4 percent on average
- Installing—65 percent on average
- Nonproductive—10 percent on average

This is an overall increase in the labor unit of 89 percent, which would certainly result in a bid that was not competitive. Clearly, it is not only essential for the estimator to understand the implications of the effect of the use of excessive manpower, but the journeymen on the job must be informed also.

Concentration. A normal quantity has been established for each of the principal classes of material which represents the quantity of material which it is reasonable to expect to find installed in a given area of a job where a journeyman would work without work moving his base of operations. When this quantity is exceeded, it is reasonable to expect a reduction in the standard labor unit. Examples of normal quantities are the following:

Item	Normal quantity	Concentration	Percent reduction
Fluorescent fixture	50	75	6
	50	144	10
Duplex receptacles	24	72	14
	24	144	17
Safety switch	4	12	13
	4	24	17

The Pricing Sheet, Form E-19, is an extremely important estimating document. Its use in conjunction with the seven work schedules when job scheduling is utilized is especially valuable, but it can also be used on small jobs where scheduling is not a consideration. Figure 6.19 shows a sample of this form, filled in with information which would normally be found in connection with the Conductor Work Schedule.

In order for Form E-19 to be used effectively, it is necessary for the estimator to flag all the deviations from standards during the process of takeoff, and then to recognize them properly on the tabulating forms; only then can they be properly accounted for on the pricing sheets. The result of careful recognition of these deviations will be seen in the ability to provide a competitive bid, which will in turn result in a profitable job.

Principle 8: Recapitulate by Work Schedule

Form E-20: Recapitulation Sheet

After completing Form E-19, Pricing Sheet, for each of the seven work schedules, the estimator has seven material cost totals and seven man-hour totals, for which it is now necessary to provide a recapitulation. In addition to those figures, the estimator may have material and labor totals for several alternate prices. Form E-20, Recapitulation Sheet, provides the means for completing this process. (See Fig. 6.20.)

Except for the alternates, the work schedules will be completed in two stages. The first stage is to tabulate the work schedules and summarize the total cost of material and total man-hours of labor. After Form E-21, Bid Summary, which is discussed in the next section, has been completed by transferring the material cost and labor hours from the Recapitulation Sheet and obtaining the percent job margin, the balance of the Recapitulation Sheet is completed to obtain a total price for all the individual work schedules and the alternates. These totals should add up to and confirm the bid price obtained on Form E-21, Bid Summary.

The main features and capabilities of Form E-20, Recapitulation Sheet, are the following:

- Material cost prices may be listed in random order by a single building, multiple single buildings, work schedules, or alternates depending on the manner in which the job takeoff is done. The companion man-hours of labor are listed in the same manner.
- Late (before bid time) changes in material prices and man-hours (if affected) are transferred from the Errata Sheet, Form E-24 (discussed in the next section), to the Recapitulation Sheet and the totals are revised accordingly so that they may be taken into consideration before summarizing the total prices.
- The completed data provided on the Recapitulation Sheet provide the basic information required to compare estimated and actual job costs by work schedule, as well as the adjusted figures required for the preparation of the Manpower Chart, Form E-23.
- When the bid alternate is taken off separately and tabulated in the same manner as a work schedule, it is posted on the Recapitulation Sheet in order to provide a completely adjusted and factored price that includes all margins and is ready to be submitted on the bid documents.

Figure 6.20 shows the Recapitulation Sheet filled in with typical information.

Form E 20

Recapitulation Sheet

JOB NAME Littleton Arts Center

SHEET NO. 1 OF 1
FLOOR NO. 1 BLDG. A
ESTIMATOR R.C.F.
DATE 4-20-84

NOTE: The form is used in conjunction with two other forms to provide completed information:

1. Form E-21, Bid Summary
2. Form E-24, Errata sheet

ESTIMATED COST OF MTL AND LABOR	539750	BID QUOTE	792,144	ESTIMATED TOTAL M-H	15,964	JOB FACTOR M-H	1,676	TOTAL SUPV.-APF M-H	3,720
LESS ERRATA ADJ.	4590	LESS ADJ COST	535,160	LESS ERRATA ADJ	136	ADJ BASIC M-H	12,648	LESS PROD SUPV.-APP M-H	2226
ADJUSTED COST OF MATERIAL AND LABOR	535,160	JOB MARGIN	256,984	ADJUSTED MAN-HOUR	15,828	PERCENT JOB FAC M-H	13.33	NON-PROD SUPV.-APP M-H	1,494
		PERCENT MARGIN						PERCENT NON-PROD M-H	11.81

ESTIMATED LABOR COST	314,480
LESS ERRATA ADJUSTMENT	1,340
ADJUSTED LABOR COST	313,140
ADJUSTED COST PER M-H	$19.87

REF. NO.	WORK SCHEDULE	MATERIAL PRICE DATA: MATERIAL COST	TOTAL SCHEDULE MTL COST	ERRATA ADJUSTM'T	ADJUSTED MATERIAL COST	PERCENT JOB MARGIN	AMOUNT MARGIN	MATERIAL SCHEDULE TOTAL PRICE	LABOR PRICE DATA: BASIC MAN HOURS	TOTAL SCHEDULE MAN-HRS	ERRATA ADJ	ADJUSTED MAN-HRS	%	JOB FACTOR MAN-HR	%	NON-PROD MAN-HRS	TOTAL SCHEDULE MAN—HRS	COST PER MAN-HRS	LABOR COST	% JOB MARGIN	AMOUNT MARGIN	LABOR SCHEDULE TOTAL PRICE	TOTAL MTL-LAB SCH PRICE
PS-1	Embedded	5506							.390														
PS-1	Embedded	2728	8234		8234	.3244	2671	10,905	112	502		502	13.33	67	11.81	59	628	19.87	12478	.3244	4048	16,526	27,481
PS-2	Feeder	22,402							385														
PS-2	Feeder	9,422							1240														
PS-2	Feeder	19,280	51,104		51,104	.3244	16,578	67,682	1335	2960		2960	"	395	"	350	3705	19.87	73,618	.3244	23882	97,500	165,182
PS-3	Equipment	58,800	58,800	-1400	57,400	.3244	18,620	76,020	1800	1800	-40	1760	"	235	"	208	2203	19.87	43,773	.3244	14,200	57,978	133,998
PS-4	Branch circuit	.3920							231														
PS-4	Branch circuit	3742							190														
PS-4	Branch circuit	1450	9112		9112	.3244	2956	12,068	196	617		617	"	82	"	73	772	19.87	15,340	.3244	4,976	20,316	32,384
PS-5	Conductor	8128							455														
PS-5	Conductor	35,932	44,060	-830	43,230	.3244	14,024	57,254	1345	1800	-26	1774	"	236	"	209	2219	19.87	44,091	.3244	14,303	58,394	115,648
PS-6	Lighting fixtures	70,000	70,000	-1000	69,000	.3244	22384	91,384	2900	2900	-70	2830	"	377	"	334	3541	19.87	70,360	.3244	22,825	93,185	184,569
PS-7	Finishing	2280	2280		2280	.3244	740	3020	120	120		120	"	16	"	14	150	19.87	2981	.3244	967	3948	6968
PS-8	Alternate #1	7400							405														
PS-8	Alternate #1	9300							302														
PS-8	Alternate #1	3880	20,580		20,580	.3244	6676	27,256	463	1170		1170	"	156	"	138	1464	19.87	29,090	.3244	9437	38,527	65,783
PS-9	Alternate #2	15,100	15,100		15,100	.3244	4898	19,998	915	915		915	"	122	"	108	1145	19.87	22,751		7380	30,131	50,129
																		ADJ	18			ADJ	10047
		TOTAL MATERIAL COST	279,270	-3,230	276,040	TOTAL MATERIAL PRICE		365,587	TOTAL BASIC MAN-HOURS	12,784	-136	12,648	TOTAL ADJ JOB MAN-HRS				15,827		314,480	TOTAL LABOR PRICE		414,615	792,144 QUOTE

Figure 6.20 Recapitulation Sheet.

On estimates for smaller jobs where subdividing the estimate into work schedules for scheduling and cost accounting purposes is not necessary, Form E-20 can be used effectively for single totals—especially where alternates may be involved.

Bid alternates. Form E-20, Recapitulation Sheet, makes it possible for the estimator to take off the work for bid alternates separately, in the same manner in which this is done on any other work schedule, and to provide a separate price for each bid alternate. In order to develop the price for a bid alternate, it is necessary to return to Form E-20 and, with the information which appears on Form E-21, Bid Summary, complete it in its entirety. It should be obvious that the bid quote developed on Form E-21 includes the price of the bid alternates.

Principle 9: Summarize the Estimate to Win the Bid

Form E-21: Bid Summary

In order to utilize Form E-21, Bid Summary, it is necessary first to complete three additional forms which supplement Form E-21 and supply information which is essential for its completion; these are:

- Form E-22, Job Factor Evaluation Sheet
- Form E-23, Manpower Chart
- Form E-24, Errata Sheet

The Bid Summary form provides a means of assembling all the costs in the estimate to arrive at a bid price that includes properly evaluated overhead and profit margins. The form includes the following sections, in which orderly determinations are made in sequence to arrive at a competitive bid price that permits a profitable job:

Estimated material cost (from Form E-20)
  Expendable material @ 0.20% labor cost

Estimated man-hours of labor (from Form E-20)

Job factor man-hours (from Form E-22)

Productive supervision man-hours (from Form E-23)

Labor cost in dollars

Direct job expense

Overhead allocation

Profit allocation

Job fees

Bid price

Errata adjustments

Form E 20, Recapitulation Sheet, provides the estimated costs of material and man-hours of labor which appear on the Bid Summary. Form E 22, Job Factor Evaluation Sheet, is the worksheet which the estimator uses to determine the values of the job factors involved. Form E-23, Manpower Chart, provides a graph from which may be determined the basic man-hours of labor, which are equal to the estimated man-hours plus the productive supervision man-hours, and Form E-23 is used to determine the total man-hours of labor in each category of labor, i.e., general foreman, foreman, journeyman, and apprentice.

Figure 6.21 illustrates the relationship between the forms that support the summarizing process, while Figure 6.22 shows Form E-21, Bid Summary, the major form in the summarizing process.

Each of the main functions dealt with in Form E-21 will now be discussed in detail.

Material cost, line 1. The total material cost figure is transferred from Form E-20 to Form E-21, and appears on line 1.

Expendable material cost, line 1a. The expendable material cost figure is determined by taking 0.20 percent of the gross labor cost which appears on line 17. While the expendable material cost is a function of labor cost, it is a part of the total material cost and is therefore added to the estimated material cost in order to arrive at a total material cost.

Estimated man-hours, line 2. The estimated labor man-hour figure is transferred from Form E-20, Recapitulation Sheet, after each of the work schedules has been posted and summarized, and appears on line 2 of the Bid Summary before job factors and nonproductive supervision have been determined on Form E-23, Manpower Chart.

Job factor man-hours, lines 3–8, and Form E-23. A *job factor* is defined as a modification to overall labor cost which is occasioned by overall job conditions that differ from the established job conditions on which the standard labor units are based. Job factors differ from the adjustment factors in that they do not apply to individual labor units; rather, they apply to the overall labor man-hours for the entire job, or for some portions thereof. The principal modifications are for the following conditions:

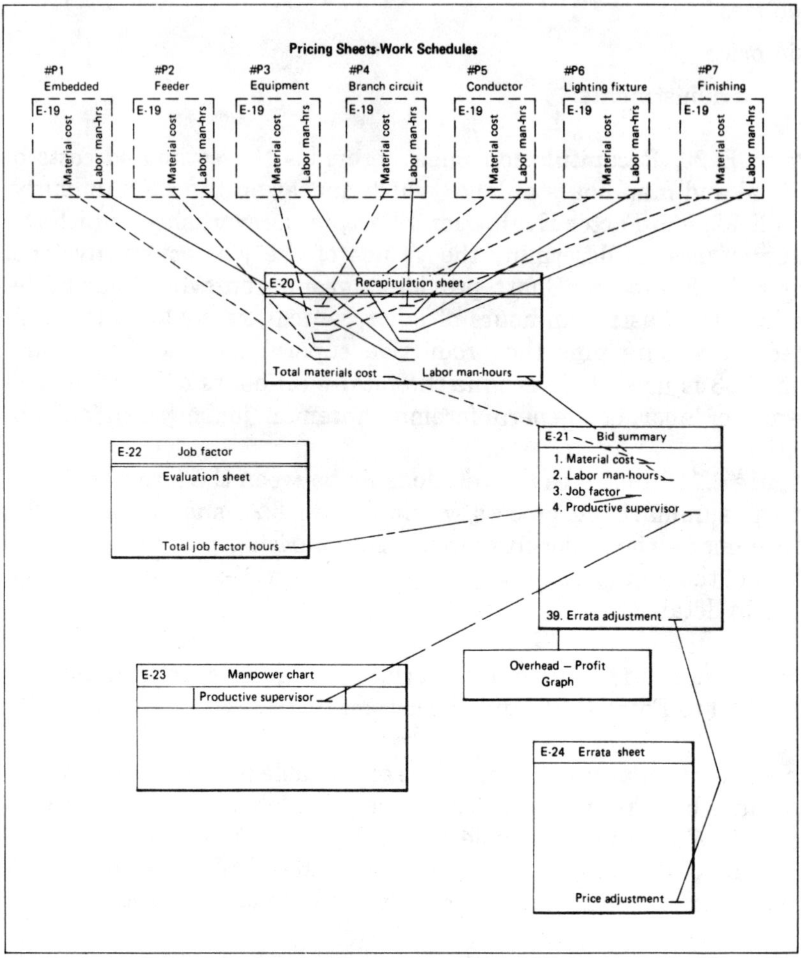

Figure 6.21 Chart of summarizing forms. This chart shows the relationships between all the summarizing forms required to furnish the information used on the Bid Summary form. It also shows the location of this information on the form.

- Height-of-building-floor variations
- Prefabrication or preassembly
- Height-of-building-riser conditions
- Productivity of manpower
- Similar buildings
- Typical floors
- Working conditions

Form E-21

Bid Summary

Sheet no. 1 of 1

Estimator R.C.P.

Job name Littleton Arts Center — Date 11/26/91

1. Estimated material cost			279270
Expendable material 0.20%	314,480		629
			279899
2. Estimated labor man-hours	Laborcost 12,784		
3. Job factors: height of building			
4. Prefabrication	− 614		
5. Productivity 9.0 %	+ 1150		
6. Similar buildings			
7. Typical floors			
8. Working conditions	+ 1150	1686	
9. Basic labor man-hours		14,470	
10. Less: productive supervision—apprentice man-hours		2226	
11. Total journeyman man-hours		12244	
12. Plus: total supervision—apprentice man-hours		3720	
13. Total labor man-hours		15964	
14. Labor cost: general foreman 880 hours @ 24.00		21120	
15. Foreman 1,440 hours @ 22.00		31680	
16. Journeyman 12,244 hours @ 20.00		244880	
17. Apprentice 1,400 hours @ 12.00		16800	314480
18. Cost of material and labor			594379
19. Direct job expense: board and lodging			
20. Equipment (trucks, trenchers)			
21. Freight on Material to the job			
22. Payroll tax 31.4 % 314480	98747		
23. Sales tax 6.25 % 279899	17494		
24. Special insurance			
25. Subcontracts			
26. Telephone expense	400		
27. Trips to the job			
28. Travel expense	800		
29. Warehouse and field office			
30. Miscellaneous			117441
31. Prime cost			711820
32. Overhead cost 15% or $6.69/Man-Hour			106773
33. Profit 4.6%			32744
34. Subtotal			851337
35. Depository fee 0.25 % 851,337		2128	
36. Performance bond 0.60% 851,337		5108	
37. Permits and inspection		4100	11336
38. Bid price			862673
39. Errata adjustment			− 4590
40. % job margin 19.60 Lab. Cost $ Man-Hour 19.70 Material 32.6 Labor 36.79 Bid quote			858083

Figure 6.22 Bid Summary form.

TABLE 6.6 Values of Job Factors

Job factor	Minimum factor multiplier	Maximum factor multiplier
Height-of-building-floor factor	+0.025	+0.30
Height-of-building-riser factor	+0.025	+0.30
Productivity-of-manpower factor	0.00	±0.25
Similar buildings factor	−0.03	−0.07
Typical floors factor	−0.056	−0.106
Working-conditions factors:		
Continuity of work	+0.10	+0.17
Conflicts with other crafts	+0.10	+0.33
Material furnished by others	0.00	+0.06
Interruptions to the work	0.00	+0.33
Rate of progress of the job by the general contractor	+0.05	+0.15
Shape of space (rooms in the building)	+0.10	+0.15
Special clothing required	+0.40	+1.00
Tooling deficiencies	0.00	+0.50
Weather of air temperature	0.00	+0.40
Work obstructions that must be cleared	0.00	+0.40

Table 6.6 lists the values of the job factor multipliers.

A separate form, Form E-22, Job Factor Evaluation Sheet, is used as a worksheet for applying these values. Form E-22 is shown in Fig. 6.23, with typical information entered. The individual job factors are discussed in detail below.

Height-of-building-floor factor. When construction work is done in a multistory building, it is necessary to factor the total labor man-hours for each floor above ground floor to compensate for the time required to transport material, tools, and workmen to the point of usage. An analysis of the labor operations required to do this indicates that an addition of 2.50 percent to the total man-hours of work for the floor above ground floor is adequate. For the second floor above ground level, the percentage adder is increased only 1.25 percent, and an additional 1.25 percent is added for each floor up to and including the sixth floor. Above the sixth floor, 2.50 percent per floor is added as shown in Table 6.7.

All the vertical conduit risers are factored using the same values as for floor heights, except that the factors for all the floors are averaged and applied to the total footage of the risers.

Prefabrication or preassembly factor. Standard labor units may be reduced when materials are prefabricated or preassembled on a shop bench or in an assembly line operation. Virtually all the eight time-consuming field operations are reduced by either of these procedures to the extent that the standard labor unit is reduced by 25 percent.

Form E-22

Job Factor Evaluation Sheet

Sheet No. 1 of 1
Estimator R.C.D.
Date 11-15-84

Job Name Littleton Arts Center

Form E-21 Line Number	Description of Job Conditions	Number of Items	Unit Quantity of Man-hours	Man-Hours Involved	Minimum Factor Multiplier	Maximum Factor Multiplier	Multiplier Used	Man-Hour Adjustment
3.	Height of building floor factor				+0.025	+0.30		
3.	Height of building riser factor				+0.025	+0.30		
4.	Preassembly or prefabrication			10,240	-0.05	-0.25	-0.06	-614
5.	Productivity of manpower			12,784	0.00	+0.25	+0.09	+1150
6.	Similar Buildings				-0.03	-0.07		
7.	Typical Floors				-0.056	-0.106		
8.	Working conditions (Itemized below)							
a.	Continuity of work				+0.10	+0.17		
b.	Conflicts with other crafts				+0.10	+0.33		
c.	Material furnished by others				0.00	+0.06		
d.	Interruptions to work				0.00	+0.33		
e.	Rate of Progress of job by G. C.			12,784	+0.05	+0.15	+0.09	+1150
f.	Shape of space (rooms or bldg)				+0.10	+0.15		
g	Special clothing required				+0.40	+1.00		
h.	Tooling deficiencies				0.00	+0.50		
i.	Weather or air temperature				0.00	+0.40	—	
j.	Work obstructions to clear				0.00	+0.40		
k.	Other							
NOTE: Line number refers to counterpart on Form E-21							Total job factor hours	+1684

Figure 6.23 Job Factor Evaluation Sheet.

TABLE 6.7 Height-of-Building Factor Values

Location	Job factor, percent*
Main floor	
Second floor and basement	2.50
Third floor	3.75
Fourth floor	5.00
Fifth floor	6.25
Sixth floor	7.50
Seventh floor	10.00
Eighth floor	12.50
Ninth floor	15.00
Tenth floor	17.50
Eleventh floor	20.00
Twelfth floor	22.50
Thirteenth floor	25.00
Fourteenth floor	27.50
Fifteenth floor	30.00

*These percentages are based on having a skip hoist available for transportation.

Productivity factor. *Productivity* is defined as "the capability of effecting or contributing to effective production." It is defined by the U.S. Army Corps of Engineers as "the capability to perform given tasks in a specified time, at a reasonable cost and with a minimum of waste." In industry it is defined as "the ratio of effective work actually done to that which could have been done under standard conditions which are uniform and well established by usage."

Lack of productivity will be manifest in the following ways:

- Substandard individual performance on the job
- Poor attitude toward performance
- Poor foreman performance in planning work
- Late starts and early stops
- Work limitations imposed by contract agreements
- Incompetence and inadequate training
- Inadequate tooling
- Two men performing a one-man operation
- Inability of the contractor to place material on the job when required
- Poor morale on the job

The degree to which a lack of productivity affects the man-hours expended on a job is largely a judgment factor, but it could vary from 0 to

25 percent. The performance of the foreman on the job probably has the largest bearing on all the factors which affect productivity negatively.

Similar buildings factor. When identical buildings are constructed and the same crew does the work in consecutive order, it is possible to reduce man-hours because of familiarity with the work. An analysis of the eight time-consuming field operations reveals that from 3 to 7 percent of total man-hours may be saved. A reduction of 3 percent is possible for the first similar building, 5 percent for the second, and 7 percent for additional buildings.

Typical floors factor. Many apartment buildings and multifloor office buildings are designed such that they will have a number of spaces with identical layouts, and this results in repetitive work by the electrical contractor's field personnel. An analysis of the operations involved results in a projected savings of 5.60 percent of the man-hours for the first repetitive layout. Table 6.8 shows the percent reduction in man-hours which applies to various numbers of typical floors.

TABLE 6.8 Values for Typical Floors Factor

Number of typical floors	Adjustment to floor man-hours, percent
One	5.60
Two	6.84
Three	8.08
Four	9.30
Five	10.60
Six	10.60
All above five	10.60

Working-conditions factor. The following working conditions affect productivity:

- Weather and air temperature extremes
- Obstructions to the work that must be cleared
- Interruptions to the work
- Conflicts with other crafts
- Material furnished by others
- Special clothing that must be worn by the electricians
- Continuity of work
- Rate of progress of the job
- Shape of the space in which work must be fitted

Each of these working conditions affects productivity, but the extent to which each is affected is a matter of judgment. The contractor makes this determination on the basis of experience with similar situations on other jobs.

Productive supervision-apprentice man-hours, line 10. Form E-23, Manpower Chart, provides a worksheet with guidelines for determining the man-hours of work projected for the job, and other important information required for good job management and accurate estimating, as follows:

- Man-hours of work for each classification of labor
- Number of man-hours which will be nonproductive for each class of labor
- Shape of the graph for manpower requirements for the duration of the job
- Manpower peaking requirements for the job
- Dates of material requirements for the different work schedules of the job
- Date of the tool and special equipment requirements

The electrical Manpower Chart must be developed around the general contractor's progress schedule, because knowledge of the duration of the job is critical to the accuracy of the Manpower Chart. However, it is difficult to persuade a general contractor to divulge his projected progress schedule before bid time because it is critical information; therefore, the entire manpower schedule must initially be developed by the electrical contractor from experience. A more accurate schedule can be made after the contract has been awarded.

Form E-23, Manpower Chart, is shown in Fig. 6.24, filled in with typical information.

The nonproductive foreman-apprentice percentage are derived from an analysis of the eight time-consuming field operations. The 40 percent general foreman nonproductivity and the 55 percent foreman are predicated on these two classifications performing productive operations which would otherwise be performed by journeymen. These operations are part of the eight time-consuming field operations and consist of studying the plans, ordering material and tools, receiving material, and storing and laying out work. These activities are all part of the basic labor units, but when performed by the supervisor they become productive man-hours. The balance of the supervisors' time is nonproductive supervision time, which is not included in the basic labor units.

Manpower Chart

Form E-23

Job Name *Littleton Arts Center*

Sheet No. *1* of *1*
Floor *1* Bldg. *A*
Estimator *R.C.P.*
Date *11-14-84*

Directions:
1. Prepare chart, and determine hours for general foreman, foreman, and apprentice
2. Determine non-productive supervision – apprentice hours
3. Determine journeyman hours
4. Recapitulate hours by labor classification

Nonproductive Hours – Supervision – Apprentice			
Title	Percent	Supv.-App. Hours	Nonprod. Hours
General Foreman	40	880	352
Foreman	55	1440	792
Apprentice	25	1400	350
Totals		3720	1494

Determination – Journeyman Hours		Man-Hours
Basic Man-Hours (Line 9, E-21)		14470
Less: Total Supervision – Apprentice Hours	3720	
Nonproductive Supervision – Apprentice Hours	1494	
Less: Total Productive Supervision – Apprentice Hours		2226
Total Journeyman Man-Hours		12244

Recapitulate Man-Hours by Title	Man-Weeks	Man-Hours
General Foreman	22	880
Foreman	36	1440
Journeyman	306	12244
Apprentice	35	1400
Totals	399	15964

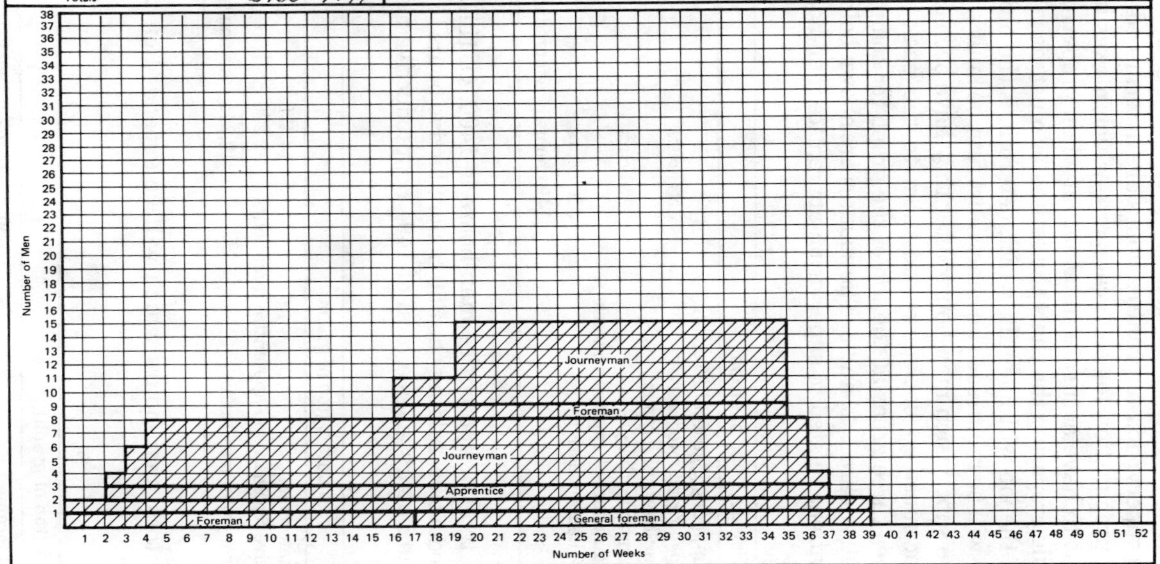

Figure 6.24 Manpower Chart.

The apprentice nonproductive time of 25 percent is an average over the 4 years of training. This percentage would be considerably greater in the first year, not so much greater in the second, and by the fourth year the apprentice is reasonably productive if used properly.

The supervision-apprentice hours are determined by the Manpower Chart preparation for each category, which makes it all the more important to have accurate information when preparing the chart.

While preparing the chart the estimator must visualize the manner in which the job will be built, which determines the order in which manpower will be required. The chart must first be built without the nonproductive supervision man-hours; after they have been determined, the estimator must go back and add them to the basic chart. Form E-23 provides three areas at the top of the graph for computation, which we will call square 1, 2, and 3. The squares are shown below to illustrate the calculations related to supervision-apprentice man-hours.

Square 1 (top of graph)	
880 man-hours of general foreman work @ 40% nonproductivity	352 hours
1440 man-hours of foreman work @ 55% nonproductivity	792 hours
1400 man-hours of apprentice work @ 25% nonproductivity	350 hours
	1494 hours of nonproductive work

The total of apprentice man-hours is 1400. Of this amount, 350 are nonproductive. Adding the 350 nonproductive hours to the 1144 supervision man-hours gives a total of 1494 nonproductive man-hours of labor.

Square 2 (top of graph)	
Basic man-hours	14,470
Productive foreman and apprentice man-hours	−2,226
Net journeyman man-hours	12,244 man-hours of productive labor

The total man-hours which will be expended on the job will be as follows:

Square 3 (top of graph)		
General foreman	880 =	22 man-weeks
Foreman	1,440 =	36 man-weeks
Journeyman	12,244 =	306 man-weeks
Apprentice	1,400 =	35 man-weeks
Total	15,964	

The graph in Form E-23 is divided into weeks; it has been found that using man-weeks is the best means of developing and plotting the graph.

After Form E-23 is completed, the figures are transferred to Form E-21. The 2226 man-hours of productive supervision-apprentice labor are entered on line 10. When this amount is subtracted from the amount on line 9, it provides the total journeyman man-hours on line 11, in this case, 12,244.

Total labor man-hours, line 13. Total labor man-hours are derived by adding the total supervision-apprentice man-hours to the total journeyman hours and posted on line 13.

Labor cost, lines 14–17. Each category of labor and the man-hour cost of each are posted on these four lines and extended to provide a total labor cost on line 17.

Cost of material and labor, line 18. This total is shown on line 18.

Direct job expense, lines 19–30. Each of these lines provides a space for consideration of an item of direct job expense, and the total is posted on line 30 to be added to the cost of material and labor to provide the prime cost.

Direct job expenses are those job costs which occur as a result of a contractor's having a particular job; if that job did not exist, there would be no expenditure of money. These costs are part of the prime cost of a job, which also includes material and labor cost. Direct job expenses are itemized on Form E-21 as follows:

- Board and lodging
- Special equipment, trucks, or trenchers
- Freight
- Payroll tax
- Sales tax
- Special insurance
- Subcontracts
- Telephone expense on or to the job
- Supervision trips to the job
- Travel expense for workmen
- Field office and warehouse
- Miscellaneous

Overhead cost, line 32. Overhead cost is that part of the cost of doing business which is fixed, and which continues every day whether or not the contractor has any jobs. The ideal situation would be to have one job which had enough margin included to meet these costs, but it should be obvious that there is no way to be competitive under a condition such as that; consequently, every job done must bear a proportionate amount of the overhead costs. Overhead has been recognized as the cost of maintaining a business in a form which enables it to provide the services required to satisfy the customers. Overhead costs include facilities, personnel, and supplies.

There are two ways to charge for overhead:

- By adding a percentage of overhead to prime cost—this percentage may be varied according to the volume of the job.
- By adding a labor burden which is based on the projected overhead cost divided by the projected man-hours of work, which provides an overhead cost per man-hour worked.

The latter system provides a much better management tool than the use of a percentage adder. At the time the budget is prepared for the year, a determination can be made as to the amount of the burden each job will carry, and this can be done for each department of the business. Hourly rates may be determined for the different types of services offered, such as service, construction, or line work. The ratio of material to labor varies considerably for different types of jobs. Using the labor burden method, there is complete assurance that the proper amount of overhead will be added for every man-hour worked, and that the total overhead will be covered if the amount of budgeted man-hours is secured in the volume of work.

Profit, line 33. It must be realized that a profit will not be earned by an electrical contracting business until enough jobs have been completed throughout the year with enough dollars included for overhead. In addition, as with overhead, some dollars for profit must be included in every job if there is to be any profit at all. It must also be recognized that the proper number of man-hours must be included in every job—otherwise, there will be no money to cover the overhead and profit because the labor cost must be covered before anything else.

Like overhead, profit should be varied according to the size of the job, and in the case of profit, the size of the job is regarded as a function of the number of man-hours involved. A graph such as that in Fig. 6.25 will enable the estimator to evaluate the job in terms of man-hours and duration in order to determine the percent of prime cost to add for profit, and the dollars per man-hour to be added for burden. The mean job size

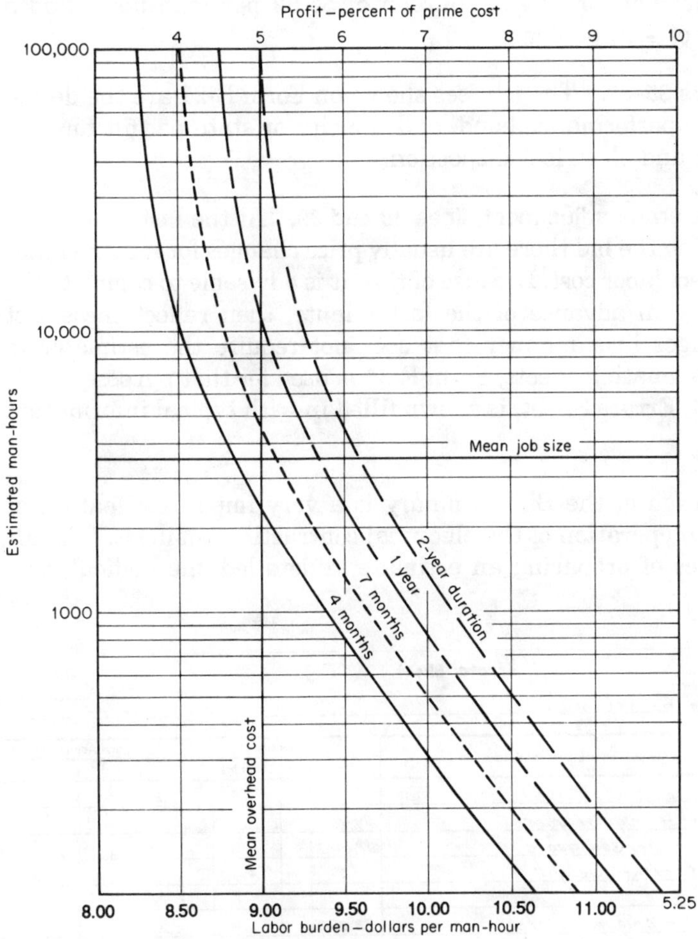

Figure 6.25 Labor-burden and profit graph.

upon which the curve is based is 3500 man-hours with a profit of 5 percent and an overhead burden of $9.00 per man-hour estimated.

Each business should prepare this graph in accordance with its own overhead costs and its own desire for profit—bearing in mind that the company must always be competitive or it will not have any business. In addition, the company must be operating in an area of the country where there is enough business activity to satisfy the volume needs of their business. The overhead-profit graph shown in Fig. 6.25 is based on a business having an annual volume of business of $4.5 million, with an overhead of $750,000 involving an average of 50 journeymen. On the other hand, the bid summary with which we have been working shows a job with a prime cost of $711,820, involving an overhead cost of 15 percent, or a labor burden of $6.69 per man-hour, and a profit adder of 4.60 percent.

Job fees, lines 35–37. The job fees shown on Form E-21 are the depository fee, the performance bond, and the city or state permit for new construction and electrical inspection.

Bid price and errata adjustment, lines 38 and 39. At the last minute before submitting the bid there are usually price changes for material that may also affect labor cost. Because of this, it is advisable to complete the Bid Summary in advance of the last minute, then reflect these last minute changes in a manner that does not require the estimator to change his estimating sheets; Form E-24 is used for that purpose.

Form E-24, Errata Sheet, is shown filled in with typical information in Fig. 6.26.

The preparation of the Bid Summary is a very important feature of the successful operation of the electrical contracting business. Like all other features of preparing an estimate, a detailed, methodical, sys-

Form E-24

Errata Sheet

Job name *Littleton Arts Center*

Estimator *R.C.P.*
Date *11/14/84*

Schedule no.	Description of revision – List by work schedule	Quantity: Material	Quantity: Labor hrs	Quantity: $	Total Adjustment: Add	Total Adjustment: Deduct
P-3	Equipment – original quote	29,400				
	Equipment – revised quote	28,000				1400
P-3	Equipment – lower weight		–40.	10		400
P-5	Conductor – original quote	22,030				
	Conductor – revised quote	21,200				830
P-5	Error in estimate		–26.	10		260
P-6	Lighting fixture – original quote	35,000				
	Lighting fixture – revised quote	34,000				1000
P-6	Fixture factory assembly		–70.	10		700
	Total labor hours		–136			
				Totals – Add or deduct		4590
				Net change		–4590

Figure 6.26 Errata Sheet.

tematic approach is necessary if the contractor is to be competitive and operate at a profit.

Principle 10: Prepare the Work for the Field

The work required to get a job started for the field is called *job processing* and is normally performed by the processor under the direction of the contract manager in the job management department. It is essential for effective operation of the business to prepare the work for the field. There are many important requirements of the job over which the field men have no control, but which have significant bearing on the job, so it is important not only to provide the answers to these questions but also to support the field on a continuing basis for smooth operations.

Essentially, job processing consists of completing all the items on the Job Processing Check Sheet shown in Fig. 6.27, and these may be categorized into five functional groups, as follows:

- Paperwork preparation and job setup
- Engineering design
- Material scheduling
- Tool scheduling
- Estimate evaluation

While there are numerous details that must be performed by the processor, the main responsibilities are the following:

- Preparing plans and specifications for field use
- Preparing the Activity Schedule for material and manpower scheduling.
- Preparing a Bill of Material with processor modifications
- Handling manufacturers' shop drawings
- Preparing tool list and requirement dates
- Preparing fixture schedule
- Preparing panelboard schedule

Preparing plans and specifications

Most of the information required on the job is conveyed to the field by plans and specifications, but additional information is nearly always required in order to provide accurate, complete, and comprehensive information. The information that is usually required to supplement the contract drawings is the following:

- Systems drawings (fire alarm, nurse's call, etc.)

Job Processing Check Sheet

Job Name ____________________ Date ________

No.	Description	Status	No.	Description	Status
1	Check job file transfer sheet		19	Prepare estimate evaluation	
2	Assign work order no.		20	Prepare cards for manpower and contract manager's desk	
3	Enter on active job list		21	Check with utility on service	
4	Enter in job log		22	Check with telephone company on service	
5	Assign contract manager		23	Check with city and county for plan approval	
6	Issue item tab to purchasing with wire comments		24	Prepare shop drawing log	
7	Prepare job file		25	Prepare material status record	
8	Apply for permit		26	Order computerized bill of material and drawings	
9	Write work order and distribute immediately		27	Prepare change order records	
10	Send group breakdowns to accounting		28	Check on introduction letter	
11	Bind specs in loose-leaf book. For field and contract manager indicate all addendas. Glue in specs. and on plans		29	Review job with estimator	
			30	Collaborate with purchasing department on proposed material list and send to gen. contractor	
12	Place office copy of plans on stick and mark		31	Prepare activity schedule	
13	Edge prints for field		32	Correct computerized bill of materials	
14	Contract original to accounting; copy to file		33	Prepare manual bills of material, include fixture cuts and shop drawings	
15	Send insurance certificates		34	Prepare tool list	
16	Request construction schedule from general contractor by letter		35	List prefab possibilities	
17	Request sales tax exemption		36	Prepare job record box	
18	Notify secretary as to file and action copy required		37	Prepare detailed drawing for field	

Figure 6.27 Job Processing Check Sheet.

- Wiring diagrams
- One-line diagrams
- Panelboard schedules
- Fixture schedules
- Equipment room layout
- Feeder conduit layout
- Busway routing
- Under-floor duct dimensional layout

Information such as this ensures adequate physical space, determines actual material requirements, and greatly facilitates installation procedures.

Note on prefabrication possibilities. As the job processing progresses, consideration must be given to the work that could possibly be prefabricated or preassembled in the shop.

The most important function of the construction department is to get to the job the right material, in the right quantity, at the right time. In order to do this, the processor has to assimilate a lot of information in a short time, since about 30 days is usually allowed the subcontractor to furnish a proposed material list to the general contractor. A form is provided for the processor to use in accomplishing this task. The purchasing agent furnishes the processor with the names of manufacturers who will furnish the materials required by the processor in accordance with the requirements of the job specifications.

The Proposed Material List is shown in Fig. 6.28.

Planning and scheduling

The Activity Schedule provides the basis for all job scheduling. An *activity* is defined as a stage of construction which must be distinguished from other stages of construction in order to ascertain that there will be no interference with the continuity of the work as a result of the unavailability of an item of material or a piece of equipment.

The value of material for an activity is not considered in the preparation of the Activity Schedule—only labor man-hours are consid-

Proposed Material List

Job name ________ Work order no. ________

Architect ________ Electrical engineer ________

Date ________ Prepared by ________ Approved by ________

Estimator ________ Job process engineer ________

Items	Specified	Proposed	Enclosed literature

Figure 6.28 Proposed Material List.

ered for each activity; therefore, the estimator must take into consideration that to permit the processor to prepare this schedule the estimate must be subdivided accordingly. The Activity Schedule must also conform to the general contractor's construction schedule.

The preparation of the Activity Schedule begins with the Activity Schedule Work Sheet. Each work schedule represents an activity, but certain items of concern within the work schedules, which usually involve embedded materials, are also activities and must be represented by an estimate of man-hours. These items may be the following: anchor bolts, under-floor duct, manhole fittings, fixture plaster frames, hanger rods, hanger racks, switchboard supports, panelboard tubs, templates, sleeves, pipe clamps.

The first step in the preparation of the Activity Schedule is the completed Work Sheet shown in Fig. 6.29. Nearly all the information except the chart is contained in this document, and as stated above, the estimate must be structured to furnish this information as required for the Work Sheet, and as found on the completed Recapitulation Sheet, Form E-20, with the total scheduled man-hours which include total supervision-apprentice man-hours.

After the Activity Schedule Work Sheet is completed, the information is transferred to the Activity Schedule, which is prepared on 1/10-in graph paper in the manner illustrated in Fig. 6.30. The activi-

Activity Schedule Work Sheet

Job: IBM—Package D Date: Nov. 18, 1984

Activity No.	Activity Description	Man-hours	Material group	Crew size	Crew weeks	Starting date	Finishing date
1	Ductlines	186	01	2	2.32	Nov 22	Dec 8
2	Embedded	462	01	3	3.85	Dec 8	Jan 4
3	Power and trench duct	965	02	4	6.03	Dec 8	Jan 20
4	Supports	360	03	4	2.25	Jan 20	Feb 4
5	Busway	252	04	3	2.10	Feb 4	Feb 21
6	Feeders	168	03	2	2.10	Feb 4	
		473		5	2.36		Mar 8
7	Cable	883	06	5	4.41	Mar 8	Apr 7
8	Equipment (owners) transformers	619	10 05	6	2.58	Apr 7	Apr 26
9	Branch circuit	2469	08	8	7.72	Jan 4	Feb 28
10	Wire	1576	09	9	4.38	Feb 28	Mar 30
11	Motor control center safety switches, wireway	516	10 04	6	2.15	Apr 26	May 11
12	Panels	418	11	6	1.74	May 11	May 23
13	Fixtures	1730	12	9	4.80	Mar 30	May 3
14	Systems	66	14	2	.82	May 3	May 4
15	Devices	154	13 & 15	2	1.92	May 9	May 23
	Total	11,297					

Figure 6.29 Activity Schedule Work Sheet.

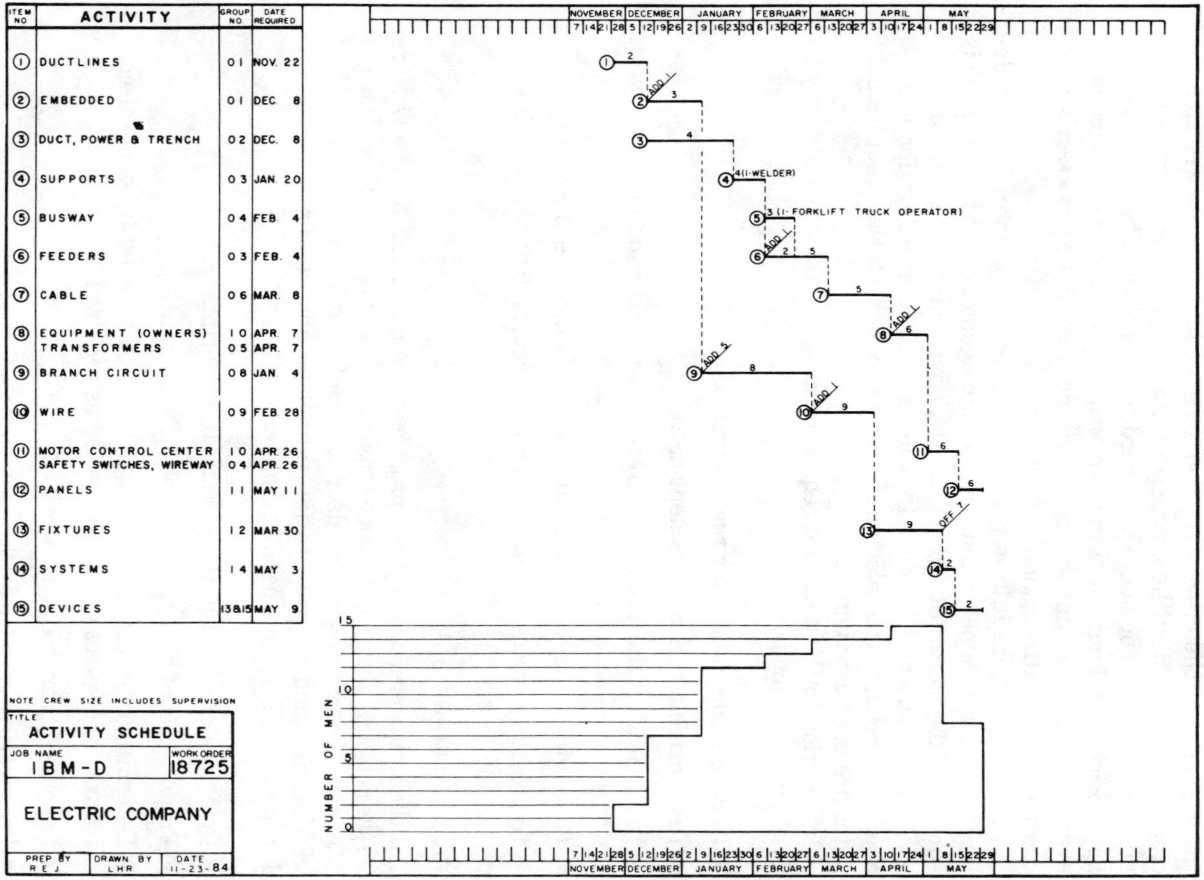

Figure 6.30 Activity Schedule.

ties are listed on the left-hand side of the sheet in a manner that allows the graph line to be projected off them. A notation is made on the graph line as to the size of the crew, and dotted lines are used to show the sequence of the crew activities. In addition, a manpower chart is created by projecting the crew size from the graph lines down to the manpower chart below. One word of caution: The supervision-apprentice man-hours will appear on the projected manpower chart evenly dispersed so that this chart will not match the processor's chart exactly, but it will, or should, be a very close match.

The Activity Schedule is the most important job management device available to supervision and top management, but its accuracy is subject to the continuity of the construction management of the general contractor, and it must be changed if the schedule for the job changes because the electrical contractor's material requirement dates are predicated on it.

In addition to the material requirement dates, the schedule may be used for the following:

- To determine manpower requirement dates
- To determine tool requirement dates
- To determine the activities involved in each subdivision of work when there is more than one subdivision
- To provide a means of determining in advance the crew size
- To provide a means of measuring crew accomplishment on site

Preparing the Bill of Material

All the items prepared by the processor are important for the welfare of the job, but the single most important document to the welfare of the overall operation of the electrical contractor's business is the Bill of Material, and its preparation cannot be undertaken until the following processes have been completed:

- Complete checkout of the electrical blueprints
- Preparation of blueprints to show feeder routing
- Determination of types of construction to be used by general contractor, and select types of assemblies required
- Layout of branch circuit work
- Preparation of fixture schedule
- Preparation of panelboard schedule
- Preparation of Activity Schedule for material requirement dates

The Bill of Material is divided according to work schedules, which in their simplest form are the following:

- Embedded work
- Feeder conduit work
- Equipment
- Branch circuit work
- Conductor
- Lighting fixtures
- Finishing

A more complete breakdown of work schedules would be the following:

- Embedded work
- Under-floor duct
- Feeder conduits
- Distribution equipment
- Transformers
- Conductors
- Switchgear
- Branch circuit conduit, fittings, and boxes
- Branch circuit wire
- Equipment
- Panelboards
- Systems
- Devices and plates

Control over scheduling will be greatly enhanced by having as many work schedules as is practical, bearing in mind that the main purpose of the work schedules is not only scheduling of material for delivery to the job, but also cost control, accomplished by having the workmen report their work time on each of the schedules for comparison to the estimated time. The work schedules must not only be practical subdivisions for easy reporting, but must also be keyed to the subdivisions of electrical construction work because the estimators must prepare cost estimates for every work schedule.

Every effort must be made to prepare a Bill of Material that is perfect in every detail, but this is very difficult for many reasons. The most important characteristics that a usable Bill of Material must have are the following:

- It indicates the right quantities.
- It gives the right catalog numbers and a good description of the items, including the color, finish, or size.
- It shows the dates required.

Five copies of the Bill of Material are made, and distributed as follows:

- Purchasing and accounting copy
- Packing ticket copy
- Contract manager's copy
- Hold copy
- Field copy

The Bill of Material is shown in Fig. 6.31.

Bill of Material Floor no. ______ Group ______ Lead time ____ Sheet no. ______

Job name ______ Ship to ______ Date ______

Contractor no. ______ Engineer ______ Contract manager ______ Work order no. ______

Quantity shipped	Quantity ordered	C L	Item code	Description	Where used	Date required	P. O. no.

Figure 6.31 Bill of Material.

Job Expense form

The services required for a job outside the company's operations are ordered on a Job Expense form such as is shown in Fig. 6.32.

This form is used in place of a Bill of Material whenever the expense is a direct job expense and is not a part of the material cost. The expense is handled in this way for accounting reasons. Typical job expense items are the following:

- Heavy hauling

Job Expense Floor no. ____ Group ________ Lead time ____ Sheet no. ____

Job name ________ Ship to ________ Date ____

Contract no. ____ Engineer ____ Contract manager ________ Work order no. ____

Description	Where used	Date required	P. O. no.

Figure 6.32 Job Expense form.

- Crane service
- Trenching
- High-potential testing
- Concrete sawing
- Core drilling
- Telephone services
- Carpentry work
- Painting
- Sound system installations

If the work is extensive and involved, a subcontract agreement should be prepared. In situations where arrangements for the work are made by telephone and a Purchase Order number given to the vendor, the Job Expense form should be prepared also, and a notation made on the form "confirming." The Job Expense form is distributed and filed in the same manner as the Bill of Material.

Fixture Schedule

It is essential for the journeyman on the job to have complete information on everything pertaining to lighting fixtures on his job. This is accomplished by preparing a Fixture Schedule such as that shown in Fig. 6.33.

The following information is shown on this form:

- Quantity of fixtures on job by type
- Information on lamps in designated fixture, including number of lamps per fixture; type, size, or shape of lamps; color of lamps; wattage of individual lamps

Fixture Schedule

Job Name ________________ Work Order No. ________________

Quantity	Type	Lamps				Description	Finish	Mounting	Manufacturer	Catalog no.	Location
		No.	Type	Color	Watts						

Figure 6.33 Fixture Schedule.

- Description of each type of fixture
- Finish of fixture
- Mounting or type of hanger
- Manufacturer's name
- Catalog number
- Area where located (names of several locations if necessary)

A horizontal line is drawn separating the fixture types after the data on each type of fixture are complete.

When the schedule is complete, it is incorporated into one of the electrical drawings for use in the field; this is done by typing the Fixture Schedule on tracing paper and cutting it into the regular tracing.

Panelboard schedule

The Panelboard Schedule is used to list the circuits and the loads they serve in any one panelboard. Figure 6.34 shows a copy of this form. The completed Panelboard Schedule contains the following information:

- Plan designation of the panelboard (LPA, LPB, etc.)
- Panelboard type
- Mounting: surface or flush

Panelboard Schedule

Job name ________________ Connected kW ________
Design kW ________

Panel ________ Type ________ Mounting ________ Voltage and phase ________

Location ________ Feed ________ In ________ Main lug location ________

Circuit no.	Breaker AMP/pole	Circuit description	Location	Watts	Circuit no.	Breaker AMP/pole	Circuit description	Location	Wat
1					2				
3					4				
5					6				

Figure 6.34 Panelboard Schedule.

- Voltage and phase
- Location
- Feeder: number, size, and type of conductors and conduit size
- Location of main lugs: top or bottom
- Connected KVA: total KVA load
- Design KVA: total KVA capacity of feeder or bus

Good design would suggest that the design KVA equal or exceed the connected KVA. Under these conditions, the demand and diversity factor would represent the spare capacity of the panelboard. Most panel manufacturers number the breakers with odd numbers on the left side and even numbers on the right side.

The Panelboard Schedule is arranged in such a way as to avoid having two circuits on the same phase, in the same conduit with a common neutral; this is because the circuit numbers in any conduit must be either all odd or all even—not a combination of both. For each circuit, the following should be shown:

- Circuit number
- Number of poles and amperage of circuit (in the case of two- or three-pole breakers, two or three circuit numbers are used, and one or two lines are blank).
- Circuit description—the load is described as the number of fixtures, receptacles, motors, etc. on that circuit
- Description of the location of the circuit area
- Approximate connected load, using 200 VA per duplex receptacle

The Panelboard Schedule is used in circuiting the job. The circuit information is listed as the circuit numbers are assigned. After the circuits have all been listed and the total connected load has been determined, the feeder may be sized.

When the Panelboard Schedules have been completed, they should be incorporated into the electrical drawings for use in the field. This is done by typing the schedules on tracing paper and cutting them into the regular tracing.

If the panelboard is connected exactly as shown on the schedule, the panelboard directory can be typed directly from the schedule. Since this procedure will save considerable time in the field, every effort should be made to connect the panelboard according to the schedule shown on the plans as prepared by the processor. The processor's work is extremely important in effecting a smooth-running operation in the field and office of the electrical contractor, and nothing should interfere with its early completion after the contract has been signed.

Principle 11: Manage the Job by Field Support

A contract manager may manage as many as 10 jobs, depending on the size and complexity of the jobs. His responsibility involves the following:

- Selection of field personnel for the jobs, including the general foreman, foreman, journeymen, and apprentices
- Supplying the job with material
- Supplying the job with tools and equipment
- Supplying the job with plans, specifications, and sketches, together with all information relative to the installation
- Handling Change Orders
- Acting as company liaison to the owner, architect, and general contractor
- Supplying the job with facilities for handling material and storing material and tools on the job
- Providing the accounting office with monthly information relative to percent completion for billing

Next to employing competent wiremen to do the work, his main task is to provide material to the job in the right quantity, and at the right time. Initially he must rely on the Bill of Material prepared by the processor, as well as all the other information relative to scheduling of material delivery and so forth, but as the job progresses and conditions change, it will be necessary for him to modify all job documents.

The overall objective of job management is to accomplish the installation of electrical work in accordance with the terms of the contract and within the framework of the estimate. It is mandatory that the work be done within the province of the prime cost—i.e., the total cost of material, labor, and job expense. In order to do this, it is essential to have a plan for effective management of the job which is disseminated to all those involved with the installation. A constant effort must be made to improve the plan, and this involves close attention to the work as the job progresses.

Material control

A major portion of the contract manager's time will probably be devoted to material control for the job—every job. Because the general contractor changes the type of construction from that shown on the plans, or because changes occur in the form of additions or deletions, the contract manager will be forced to change the Bill of Material by revision, addition, or deletion. His success, or lack of success, in controlling material is manifested in three ways, as follows:

- The amount of material ordered from the field during the job
- The amount of material returned to the shop at the end of the job
- The number of delays or interruptions in the actual work assignment due to lack of material or improper material (this measure of failure will probably never be known)

If the Bills of Material were perfect none of the aforementioned problems would occur—providing the scheduling was perfect, the purchasing department functioned as it should, and the suppliers met their obligations as required.

Changes to the Bill of Material

Changes are accomplished in the following manner:

- Revisions are accomplished using the form shown in Fig. 6.35, which is titled Revision to Bill of Material.
- A material scheduling change is made on a form of that name, shown in Fig. 6.36. This is usually a change in date required.
- An additional Bill of Material may be prepared.

Material scheduling changes are usually the result of changing the Activity Schedule when the progress of the job is accelerated or falls behind. This is serious because it may affect a number of activities. Nevertheless, the Activity Schedule should be reviewed frequently and appropriate changes made when required.

Selection of personnel to work in the field

The success of a job depends to a large extent on productivity of the work force, which is largely a function of the competence of the field workmen; therefore, the selection of all the workers in the field is very important. Once competent personnel have been selected, it is the contractor's responsibility to provide maximum field support. No matter how conscientious the field man might be, there are many things that relate to the success of the job over which the field man has little or no control, but which lie within the contractor's control. That labor is the foremost ingredient in installation work is undeniable, but if productive man-hours of work are to be attained it is mandatory for the con-

Revision to Bill of Material

Job Name ______________________ Date __________

Work Order No. __________ B/M No. __________ P. O. No. __________

Quantity | Description | Change From | To

Reason for Change ______________________

____

____

______________________ Contract Manager

______________________ Purchasing Acknowledgement

Figure 6.35 Revision to Bill of Material.

Material Scheduling Change

Job Name ______________________ Work Order No. __________

Change 'Date Required' on Following Bills of Material as Indicated:

Group No.	Description	Purchase Order Number	Bill of Material Number	Date Required Changed From	To

Signed ______________________ Date __________

Note: Purchasing is to notify the Contract Manager immediately if the above revised dates will not be met, stating reasons.

Figure 6.36 Material Scheduling Change.

tractor to control the flow to the job of the right kind of material, in the right quantity, and at the right time. *Productivity* may be defined as "the capability of effecting or contributing to effective production." Therefore, the capability as defined by the competence of the workmen is a primary ingredient in productivity.

Supply of tools to the field

It is essential for jobs to be equipped with up-to-date tools of every kind. Equipping the job is only one step in the overall provision of tools; the others are:

- Tool purchase and rental
- Tool storage
- Tool maintenance and repair
- Tool retirement
- Tool records
- Tool disbursement and return

All the foregoing functions require special care and treatment, but the workers in the field are expected to take good care of all tools by doing the following:

- Using tools properly
- Performing proper field maintenance and storage
- Tagging all substandard tools and returning them to the shop for service

Regardless of the way tools are used and cared for in the field, it is essential that all tools sent to the field be in excellent working condition, and that all parts mandatory for the operation of the tool accompany it to the field. When a tool is in need of service, the field man must fill out Tool Maintenance and Repair form shown in Fig. 6.37, and return it to the shop with the tool.

Supply of information to the field

At the beginning of the job, and as the job progresses, all the information pertaining to the job will be provided to the field and maintained in the six-pocket job file, as follows:

- Pocket 1—general information
  - Job Information Sheet

Tool Maintenance and Repair

	Tool parts missing or damaged	Dull cutting edges	Supplies missing or inadequate	Operation electrical
Bag, screw			Replenish stock	
Bender, hydraulic	Shoes, pins, bars decals, etc.		1 pt of hyd. oil	Test-pump motor
Box, cable-splicing			Replenish stock	
Box, gang	Hinges, hasp doors, rollers			
Can, oil-cutting or lube			Replenish oil	
Carts	Wheels			
Cord, extension	Caps and plugs			Test
Crimper, hydraulic (hypress)	Dies			
Drill, core	Adaptors			Test
Drill, core sump pump				Test
Drill, core barnacle	Gasket			
Drills	Chuck key			Test
Drill, press	Chuck key			Test
Drill, press magnetic	Chuck key			Test
Drive power	Chuck			Test
Fastener, powder	Pins, etc.			
Generator			Gasoline	Test
Grinder, bench			Emery wheels	Test
Hammer, electric	Chisels, drills		Grease	Test
Hammer, drill phillips		Percussion bits	2¼ in Percussion bits	Test
Hammer, drill (large-medium)	Drills			Test
Hose, garden	Connections		Gaskets	
Ladder	Steps, rungs sides			
Levels	Broken vials			
Puller, cable motorized	Cable			Test
Puller, cable manual	Cable			
Punch, KO hydraulic	Cutters, bolts	Cutters		
Punch, KO ratchet	Cutters, bolts	cutters		
Saws, all kinds			2 extra blades	Test
Tape, fish vacuum	Parts			Test
Threaders	Dies	Dies		

Figure 6.37 Tool Maintenance and Repair form.

- Manpower Charts
- Activity Schedule
- Estimate evaluation
- Permit
- Material Status Report
- Notes to the contract manager
- Notes and questions for the general contractor and architect
- Job File Transfer Check Sheet
- Job Processing Check Sheet

- Pocket 2—contract data
  - Contract
  - Bid Summary

- Work Order
- Divisions of job in estimate
- Contract Change-Order Record; Time and Material Change-Order Record
- Proposals

- Pocket 3
  - Bills of Material
  - Receiving and inspection reports
- Pocket 4
  - Sketches
  - Manufacturers' shop drawings
- Pocket 5
  - Main incoming correspondence
  - Main outgoing correspondence
  - Miscellaneous correspondence
  - Interoffice correspondence
- Pocket 6
  - Estimator's worksheets
  - Interoffice mail
  - Library of technical books and manufacturers' catalogs
  - Manpower directory
  - Records, files, logs, and tool inventory
  - Reproductions of photos
  - Office supplies such as pencils, erasers, forms, paper clips.

Job visitations

Job visitations by the contract manager are a very important part of supporting the field; they are made on a regular basis for the following reasons:

1. To determine whether or not the field has been supplied with the right kind of material, tools, manpower, and information in the right amount and at the right time.
2. To determine whether or not the field is using these items as intended.
3. To determine whether or not the field is working with maximum productivity.

Item 1. Before top performance can be expected from the field they must be furnished the necessary ingredients of the job; this should be done in such a way that the actual installation of material is about all that the field personnel need to be concerned with. If the contract manager's efforts have been effective, the field will not have been handicapped by the following:

- Having to determine what material to order from the shop
- Having to lose time waiting for material from the shop
- Receiving the wrong material from the shop
- Receiving tools in unworkable condition
- Waiting for information from the office

If the job is burdened by any of the foregoing problems, it is the responsibility of the contract manager to eliminate them promptly.

Item 2. The contract manager must do the following to ensure that the items that are so necessary in the field are being used properly:

- Material
  - Check requirement dates on Bills of Material against immediate need and delivery dates shown on the Activity Schedule.
  - Plan the work with the job foreman as far in advance as practical and make sure that material and tools are either on hand or available for the next work activity.
  - Ascertain whether all material-handling devices are being properly used.
  - Ascertain whether Bills of Material are being properly filed.
  - Determine whether or not material is being received and stored properly.
- Tools
  - Ascertain that all tools are working at top efficiency.
  - Make certain that all tools are in use and not being stored.
  - Determine whether or not tools are being properly cared for.
- Manpower
  - Check out the current Manpower Distribution Report.
  - Make certain that the foreman is running his crew and performing the productive time-consuming tasks other than installation.
  - Ascertain that one-man tasks are being performed by only one man.
  - Determine that starting and quitting times are being properly observed.
  - Monitor coffee break times.
  - Look for potential trouble spots, as follows: professional jealousies, conflicting personalities, cliques, labor lawyers, nonproductive workmen, and unsafe practices.

- Information
  - Make certain that revised drawings are available.
  - Make certain that wiring diagrams are adequate, and no design work is being done in the field.
  - Determine that all Change Orders are being properly handled.
  - Ascertain that all Time and Material Change Orders are turned in to the shop as soon as they are completed.
  - Encourage the foreman to be a good housekeeper.

Most field personnel are excellent mechanics, but they seem to inherently resist paperwork, and postpone it as long as possible. This means that they must be encouraged to process their paperwork as quickly as possible, and keep it up to date.

Since the contract manager is not on the job all the time, it is essential for him to instill enthusiasm in the job foreman, and motivate him to instill it in the whole crew. It is important for him to gain the confidence of the foreman so that the foreman can share the responsibility of determining the crew size, the placement of additional foremen, and the termination of substandard wiremen.

Item 3. It is of great value to management to know how a job is progressing in terms of man-hours estimated versus man-hours actually worked, taking into consideration the actual percent complete. Weekly reports show the man-hours estimated for each work schedule, and they also show the field man-hours for each category of work completed up to that date. In order to determine the progress of the job as of that date, the contract manager must determine the percent complete for labor expended for each work schedule. This determination is made by surveying the job carefully and estimating the percent complete. When this has been done, the estimated percent completed may be compared with the percent complete as shown by comparing the man-hours.

If the figures reveal a deficiency in the progress of the job, remedial measures must be taken to improve productivity.

Manufacturers' shop drawings

Most jobs require some electrical equipment of a special nature such as custom-made equipment for a special use; for this the manufacturer is required to furnish to the electrical contractor shop drawings for approval of the owner. A Shop Drawing Log, shown in Fig. 6.38, provides a systematic method of recording the steps required to obtain this approval before proceeding with the manufacture, so that the equipment will be available and on the job at the proper time. The Shop Drawing Trans-

Shop Drawing Log

No. of copies required ____________ (X+5)

Job name ______________________ Date ____________

Min 2 weeks Max 2 weeks

Item	Manu-facturer	Bill of material	P. O.	Date ordered	Date required	Date from purchase	Date preliminary to field (A)	Date to general contractor (B)	Date from general contractor	Action A approved A/C approved as corrected D/R disapproved resubmit	Date to purchase (C)	Date approved copies to field (D)

Figure 6.38 Shop Drawing Log.

mittal form, shown in Fig. 6.39, is used in transmitting shop drawings to the field, to the general contractor, and to the purchasing agent.

Use of the contractor's shop by the field

Many tasks required in any electrical installation can be accomplished more satisfactorily in the shop than in the field. Most of the work that is performed in the shop for the field is either prefabricated, preassembled, or precut. The use of the shop by the field is in fact almost unlimited. The following operations can be done very effectively in the shop, and much better than in the field:

- Cutting, bending, and threading special sizes and quantity of conduit
- Assembling and prewiring fixtures, control units, and racks of safety switches
- Fabricating special boxes
- Quantity drilling and tapping
- Welding of all types
- Precutting under-floor duct
- Fabricating flexible conduit for laying fixtures
- Performing special assembly operations
- Operations that require special tools, jigs, or templates

Cable-handling procedures

The estimator takes off the feeder runs with the confirmation of the processor, and the required footage of cable is shown on the Bill of Ma-

Shop Drawing Transmittal

To: ______________________ Date: ______________________

______________________ Bill of material no. ______________________

______________________ P. O. No. ______________________

Attention: ______________________ Electrical contractor transmittal no. ______________________

Job name: ______________________ General contractor transmittal no. ______________________

Gentlemen:

Attached herewith are the following ship drawings as prepared by

______________________ ______________________

______________________ ______________________

which have been:

☐ Prepared for your approval ☐ Approved ☐ Disapproved, resubmit

☐ Corrected and/or revised for your approval ☐ Approved as noted ☐ Prepared for your records These are FINAL

No. of copies	Drawing number	Description	Date of drawing

☐ Release for shipment as scheduled

☐ Please return three (3) "APPROVED" copies

____

Figure 6.39 Shop Drawing Transmittal form.

terial in total footage. It is hazardous to cut the cable and reel it for usage on the job without field-measuring the cable beforehand.

Another very important consideration when purchasing the cable for the job is the estimated price in the contract. Sometimes a whole year expires between the time of the estimate quote and the time at which the job is ready for the cable to be pulled. In a volatile copper or aluminum market the price of wire can change appreciably. In order to protect the estimate price, a commitment must be made to purchase the cable at that price, with the understanding that the cable will be cut and reeled according to field measure. Therefore, a precise cable-handling procedure must be followed in the operation, as follows:

- A Bill of Material is prepared by the processor showing a feeder listing similar to that shown in Fig. 6.40, the Feeder Cutting Lengths

Feeder Cutting Lengths

Job name Empire State Building Work Order No. 18971

Date requested ______ Date originally scheduled ______

Feeder no.	No. of pcs.	Size	Insulation		Conduit material	Length	Packaged	B/M	P. O.	Date required
			Type	Color						
LPA	3	3/0	THW	Blk.	A1	98 ft	3 reels	28972	39056	10/31
PP	3	3/0	THW	Blk.	A1.	90 ft	3 reels	28972	39056	10/31
PP	1	3/0	THW	Blk.	A1.	87 ft	1 reel	28972	39056	10/31
Service	3	250	THW	Blk.	A1.	132 ft	3 reels	28972	39056	10/31
Service		250	THW	Blk.	A1.	128 ft	1 reel	28972	39056	10/31
Elevator	3	2	THW	Blk.	Cu.	73 ft	3 coils	28972	39056	10/31
Elevator	1	2	THW	Wht.	Cu.	78 ft	1 coil	28972	39056	10/31

Figure 6.40 Feeder Cutting Lengths form.

form; in the Bill of Material, each feeder is identified by plan number and described in the Where Used column.

- Requirement dates are established for both the regular Bill of Material and the feeder listing.
- Orders are placed by purchasing according to the Bill of Material, with the understanding that quantities will depend on field measure at the time of requirement.
- When the exact cutting lengths have been determined by the field, they are transmitted to the purchasing agent through the contract manager on the standard Bill of Material as shown in Fig. 6.41.

Systems diagrams

For best results in the field, it is essential that complete and comprehensive diagrams of modern signal and control systems be made expressly for the particular job. These diagrams should consist of the following:

- A schematic (ladder) or elementary diagram showing the overall function of the system, including all the control components and devices without regard to relative physical position.
- A complete wiring diagram showing the size of conduit and conduc-

Bill of Material Floor no. ______ Group (Cable field measure) ______ Lead time ______ Sheet no. 6a

Job name Empire State Building Ship to Job-site Date ______

Contr. no. ______ Engineer ______ Contract manager V.T. Work order no. 18971

Quantity shipped	Quantity ordered	C L	Item code	Description		Where used	Date Req'd.	P.O. no.
	3 pcs.		330423	Wire, A1. 3/0, THW Black	98 ft	LPA	10-31	
				(3 – reels)				
	3 pcs.		330423	Wire, A1. 3/0, THW Black	90 ft	PP		39056
	1 pc.		330423	Wire, A1. 3/0 THW Black	87 ft	PP		
				(4 – reels)				
	3 pcs.		330425	Wire, A1. 250 MCM, THW Black	132 ft	Service		
	1 pc.		330425	Wire, A1. 250 MCM, THW Black	128 ft	Service		
				(4 – reels)				
	3 pcs.		331402	Wire, Cu. #2, THW, Black	73 ft	Elevator		
	1 pc.		331402	Wire, Cu. #2, THW White	78 ft	Elevator		
				4 – coils)				

Figure 6.41 Bill of Material (cable field measure).

tors and the number of conductors to be installed from component to component. The conductors should be identified by color or number, and it should be shown where each conductor is to be terminated.

The schematic diagram is essential to understanding the function of the system and is very helpful in locating trouble or in making changes in the system. It is first made as part of the process of designing the system, and, of course, it is the base from which the wiring diagrams are made.

For some jobs it may be necessary to make the wiring diagrams on two drawings. One drawing will show the conduit and wire, and the other will show the terminations. The system can be "piped" and "pulled" from one drawing, and terminated from the other. Both functions are important from the standpoint of the electrical contractor, but it would be very time-consuming to have to make the terminations without the "termination" drawing, using only the conduit and wire drawings. Therefore, it is most essential for the termination drawings to be available to the field, because it is on these drawings that the terminal identification of one component is listed opposite the terminal designation of the other, with the connecting conductor identification in between.

Communications

The importance of communications between individuals attempting to coordinate their efforts to accomplish a common goal cannot be overemphasized. Most communications between individuals are accomplished by oral or written means. Oral communication has its place, but the advantages of the written are pronounced. The different means of communication are the following:

Face-to-face communications are essential in personal relationships, but when important conclusions are reached they should be confirmed in writing.

Telephone communications are essential and a good telephone personality should be aspired to. Words and voice inflections are all the caller has with which to evaluate the conversation, so great care must be taken to recognize this.

Written communications involve the contract, letters, standard procedures and forms, memorandums, and field notes.

Job facilities for material and tools

Productivity is greatly enhanced by providing the field with good facilities for storing material, tools, and equipment. A vast majority of field men are not particularly orderly and systematic about the manner in which they care for the contractor's belongings on the job, so it is imperative to provide them with unusually good facilities to do so. These facilities consist of tool lockboxes, material storage shelves, tote boxes, carts, two-wheel trunk carriers, four-wheel heavy-duty carts, plan desks, 50-ft trailers, and so forth. With these facilities available there is every reason to expect the workers to maintain an orderly situation on the job, and this definitely contributes to improved productivity.

There is absolutely no substitute for willing and dependable performance on the part of the workmen in the field, but there is also every reason for the electrical contractor to provide maximum support on an ongoing basis.

Principle 12: Finalize the Completed Job

The importance of closing out the job properly cannot be overemphasized. There are four principal reports that must be completed as the job comes to a close. These are:

- Job Closeout Checklist
- Report on the results of the job

- Job Evaluation Report
- Estimate Evaluation Report

It is very important that all jobs be closed out as quickly as possible after the work has been completed. Failure to do this results in a delay in collecting the final payment and the retainage, which usually amounts to 10 or 15 percent of the total contract price.

On most construction jobs, many man-hours are wasted during the final stages of construction because of failure to account for all minor items of material required to complete the job. End-of-job waste of man-hours may be caused by the following specific problems:

- Material shortages
- The need to replace fixture glass, box covers, panel trims, switch and receptacle plates, etc.
- Specialty items like systems
- Damaged lighting fixtures
- Final connections to late-arrival equipment
- Punch list items and inspection reports
- Panel directories
- Final decisions by the architect or owner on questionable items
- Waiting for other crafts to clear conflicts
- "Fine print" items on plans and in specifications

For most jobs, there is very little allowance made on the Bid Summary sheet for problems such as the foregoing; they should therefore be dealt with as the work progresses and not allowed to drag out to the last minute as a special item.

The Job Closeout Checklist is shown in Fig. 6.42. It is especially important to fill out this form well in advance of the completion date because it can be quite time-consuming to complete.

The report on the results of the job is a written report which is prepared by the contract manager; it is intended to give his overall impression of all important aspects of the job. The Job Evaluation Report is prepared by the construction manager after he receives the written report on the results of the job from the contract manager. A copy of the Job Evaluation Report form is shown in Fig. 6.43.

The Job Evaluation Report is the result of accounting for differences between the estimated amounts and the actual dollars spent. In order to do this satisfactorily, it is necessary for data to be accumulated dur-

Job Close-Out Checklist

Job Name ______________ Work Order No. ______

Contract Manager ________ Job Foreman ________ Date ______

Field Work	Date Completed
1. Conduct Prefinal Inspection to be made by Contract Manager with Job Foreman prior to final inspection by the Architect or City, County, or State Inspector. Check for the following items: a. General compliance with plans and specifications. b. Covers, plates, panel trims, plastic or glass parts DAMAGED OR MISSING. c. Panel directories (typed and in place) d. Trash removal and General cleanup e. Company name plates on panels, switchgear, etc.	
2. Call for final inspection by City, County or State Inspector. Make certain inspector signs the permit card on the job.	
3. Complete all punch list or deficiency items.	
4. Ready tools, material, trailer and/or equipment for pickup. Be sure yellow tags are on all defective or inoperative tools.	
5. Indoctrinate operating personnel as required by the specifications.	
6. Deliver the following items as required by the specifications: a. Spare fuses b. Keys to panels c. Spare parts GET SIGNED RECEIPT FOR ABOVE ITEMS	
Office Work	
1. Write memo to Sales Manager requesting a Salesman be assigned this customer and recommend a time to visit the job with the Salesman to introduce him to the proper customer's Representative. This should be done at least one week before our men are off the job.	
2. At least one week before men will be off the job, obtain a complete list of tools still charged to the job from the tool man and give to the Job Foreman for an accounting. All tools must be accounted for in one way or another.	

	Date Completed
3. Transmit the following items as required by the specifications: a. "As Built" drawings b. Operation and maintenance manuals c. Cost data, spare parts list, etc.	
4. Write letter as required by the specifications, confirming the guarantee requirements (use the same wording as in the specifications). Cite the date acceptance or beneficial occupancy from which the year's guarantee is to be determined. If a date has not been formally established – establish one.	
5. Ascertain that all field copies of all work orders and change orders are completed, signed, dated, and transmitted to accounting.	
6. Ascertain that all required sales tax and payroll documents have been issued by accounting.	
7. Ascertain that all FINAL billings have been rendered.	
8. Remove job from the "Active Job List."	
9. Issue Close-Out Ticket to Warehouse	
10. Write a narrative explaining the outcome of the job. Transmit to Construction Manager. Include Job Close-Out Check Sheet, and Estimate Evaluation Sheets.	
11. After FINAL payment is received and ALL job responsibilities have been completed, DESTROY all papers in the Job File except the following: a. One complete set of latest plans b. One complete set of specifications and addendas c. All correspondence (both incoming and outgoing) d. One set of approved shop drawings e. Any other documents deemed worth saving Transmit these items to the draftsman for proper storage.	
12. Write letter of appreciation to Foreman, if warranted.	

REMARKS: ______________

Signed ______________

Figure 6.42 Job Closeout Checklist.

Job Evaluation Report

Job Name ______________________ Work Order No. ______________

Salesman ______________ Estimator ______________ Processor ______________

Contract Manager ______________ Foreman ______________ General Contractor ______________

	Estimated		Actual	Percent of Final Estimated	Percent of Change
	Original	Final			
Material	$	$	$		
Labor	$	$	$		
Job Expense	$	$	$		
Prime Cost	$	$	$		
Margin in $	$	$	$		
Margin in %	%	%			
Total Contract	$	$		____	____

CONTRACT MANAGER'S EXPLANATION OF RESULTS: Use facts, figures, opinions.
Attach accumulated Estimate Evaluation Reports, Group Labor Reports.

DATE ______________ SIGNED ______________________
(USE ADDITIONAL SHEETS IF NECESSARY)

Figure 6.43 Job Evaluation Report.

ing the entire course of the job, but in general these differences occur because of the following:

- Errors in the estimate as a result of
  - Faulty arithmetic
  - Incorrect quantities
  - Incorrect types of material
  - Incorrect labor unit application
- Failure to purchase according to estimate
- Failure to deliver material and equipment as scheduled
- Failure to manage the job properly because of
  - Inefficient manpower
  - Inadequate or incorrect information

- Inefficient methods and tooling
- Poor and inadequate planning and scheduling
- Inadequate and ineffective supervision

- Circumstances beyond the contractor's control
  - Weather conditions
  - The general contractor's performance
  - Theft of materials

When an error is discovered in an estimate, it should be reported immediately in an effort to assist in improving estimating accuracy, as well as to aid in accounting for any overall differences between estimated amounts and the actual job cost.

Estimate Evaluation Report. The Estimate Evaluation Report form, shown in Fig. 6.44, is used at any time during the progress of the job to account for errors in various forms of estimating which should be corrected. In addition to the items on the form which are self-explanatory, it may be used for the following purposes:

- It must show the percentage completed of the job.
- A complete description of the error, and how it should have been handled, should be given.
- The costs of material and labor may be approximated.
- It must be indicated whether the error is a loss, an omission, or an excess with a positive connotation.

It is important to note that an excess is an error as much as a loss is, and it is important to recognize it as such.

Estimate Evaluation Report Report No.______ Job____ %

Date__________

Job Name________________ Work Order No.________

Estimator______ Salesman______ Processor______ Contract Mgr.______

No.	Description	By	Material	Labor	Total	+ or -
			$	$	$	

Figure 6.44 Estimate Evaluation Report.

Completing a job and closing it out expeditiously may save it from being a loser, and can prevent many disagreeable moments for the owner, architect, general contractor, and above all the electrical contractor himself. Job closeout must therefore be done in a methodical manner in order to ensure that it will be an effective operation.

Principle 13: Manage Material for Job Productivity

In previous portions of this book we have shown how to lay the groundwork that makes it possible for the purchasing department and warehouse of the electrical contractor's operation to make available for the job, material of the right type, in the right quantity, at the right time. We have stated several times before that this is the whole responsibility of the contractor in helping to make the job function properly to achieve maximum productivity on a job which the contractor has bid in a competitive manner. These things just don't happen without having a lot of well-trained people do the things that make them possible. All the things done by different individuals must be coordinated to result in a body of work which makes effective material management possible. Needless to say, the procedures that result in having material, equipment, tools, and facilities available when required are only a part of a complete business system, but they are a part that is indispensable to the success of the job.

Some contractors operate without using a purchasing agent to accomplish the measures which will be discussed in this chapter, but however they may be accomplished, the measures proposed here must be taken to achieve maximum productivity and outstanding service to the construction industry. Purchasing must look to estimating and job management to provide the information in the form in which it is needed, and in the time in which it is required, in order to obtain the material for the warehouse to deliver in proper containers according to schedule.

At the time of estimating, all the quotations for material prices are obtained by and issued by the purchasing department. The estimator prepares a carbon copy of the Pricing Sheet and transmits it to purchasing for quotations. When prices are received, they are noted in the proper place and returned to the estimator for use on the original Pricing Sheet. When the job is awarded, all the quotation sheets are sent back to purchasing so that the process does not have to be repeated.

The following processes are involved in the purchasing functions as they relate to material management:

- Purchase Order procedure
- Definition of expendable material

- Job expense procedure
- The Bill of Material
- Revisions to Bills of Material
- Conductor-handling procedure
- Expediting procedure
- Purchasing and Expediting Report

The following processes are involved in the warehousing functions as they relate to material management:

- Material storage and disbursement
- Shop Delivery Order procedure
- Receiving and Inspection Reports
- Job facilities for handling and storing material
- Tool storage and repair
- Job facilities for handling and storing tools

Purchasing policy for construction jobs

The Purchase Orders for all items of factory-ordered material for each job should be placed as soon as possible for delivery in accordance with scheduled dates issued by the processor. All work schedule materials are processed by the expediting clerk after an appropriate lead time has been established for each group. From this information, the Material Schedule Report shown in Fig. 6.45 is prepared, which results in

Material Schedule Report

Requirement Date ____________________

Lead Time ____________________

Class ____________________

P. O. Number ____________________

Job Name ____________________ Work Order No. __________

Floor Number ____________________

Contract Manager Number ____________________

Date Transmitted to Accounting ____________________ Time __________

Figure 6.45 Material Schedule Report.

the issuance of the Purchasing and Expediting Report shown in Fig. 6.46. All groups of material scheduled for initial release to the job are disbursed by the warehouse without a scheduled date, while all the others are set up on a requirement date.

While the contract manager is responsible for scheduling and verifying requirement dates for the material delivery, it is the purchasing department's responsibility to place the material on the job at the designated time.

Fundamental to all procedures in the purchasing department is the proper preparation of the Purchase Order, showing all the information

Purchasing and Expediting Report 5 5/ 7/82

Work order	Contract mgr.	Class	Floor	Purchase order	Lead time weeks	Required date	Revised requirements date
18066-00	50	A		12460	8	7/01/82	
18066-00	50	H	4		2	6/01/82	
18066-00	50	R	4		2	6/01/82	
18066-00	50	C	4		2	6/01/82	
18066-00	50	M	4		2	6/01/82	
18066-00	50	R	2		2	6/01/82	
18066-00	50	H	2		2	6/01/82	
18066-00	50	C	2		2	6/01/82	
18066-00	50	X	7		2	6/01/82	
18066-00	50	W	7		2	6/01/82	
18066-00	50	F	7		2	6/01/82	
18066-00	50	D	7		2	6/01/82	
18066-00	50	I	6		2	6/01/82	
18066-00	50	H	6		2	6/01/82	
18066-00	50	R	6		2	6/01/82	
18066-00	50	C	6		2	6/01/82	
18066-00	50	M	6		2	6/01/82	
18066-00	50	I	5		2	6/01/82	
18066-00	50	H	5		2	6/01/82	
18066-00	50	R	5		2	6/01/82	
18066-00	50	C	5		2	6/01/82	
18066-00	50	R	2		2	6/01/82	
18066-00	50	H	2		2	6/01/82	

Figure 6.46 Purchasing and Expediting Report.

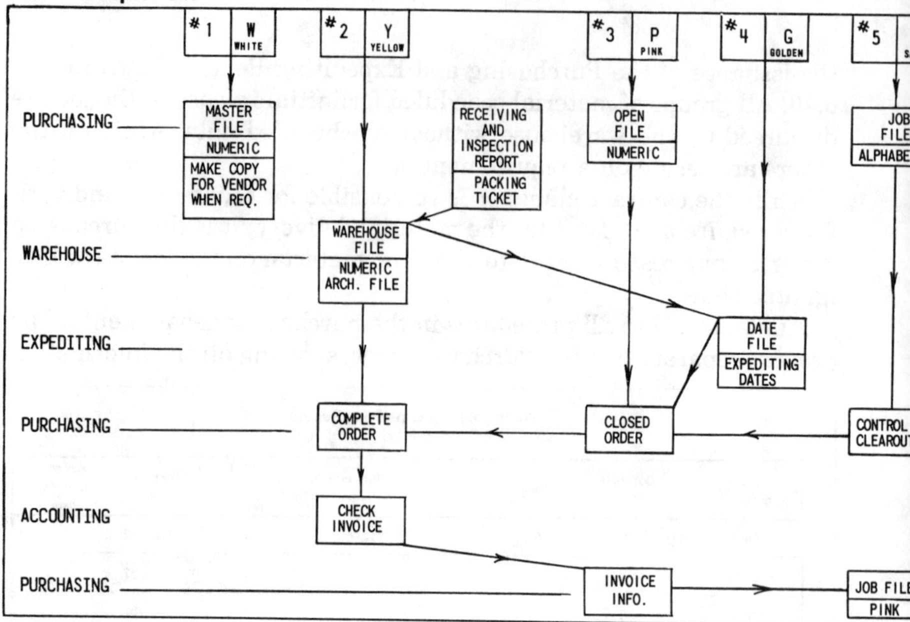

Figure 6.47 Purchase Order flowchart.

essential for the smooth performance of material handling. A Purchase Order flowchart is shown in Fig. 6.47.

The following information must appear on the Purchase Order:

- *Show in heading:* Purchase Order number, vendor's name and address, date
- *Show in body of form:* quantity, description, shipping date, cost, requirement date, how shipped, ROB point, Bill of Material number, terms, job name, Work Order number, account number, delivery point.
- Signature of buyer and shop drawing schedule

The following are important considerations in regard to Purchase Orders:

- The order form must be completely filled out.
- Separate orders should be written for each requirement date.
- Back orders must be avoided, even if the order must be placed with multiple vendors.
- Separate Purchase Orders should be written for the items on each work schedule.
- At least two quotes must be obtained, and the price paid must conform to that used in the estimate.
- Purchasing must meet requirement dates.

Procedure for handling Change-Order material

Most Change-Order material requires immediate processing, and is shipped in its entirety from the company warehouse for immediate delivery. Material to be used on Time and Material Change Orders is handled similarly to that on a Bill of Material which is prepared by the contract manager.

Expendable materials

Expendable material is defined as that which is necessary for the installation work, but for the most part does not remain as an integral part of the structure. These items do not appear on the Bill of Material, but they are carried in inventory and charged to the job like any other item of material on a material order. Inasmuch as they cannot be estimated like any other item of material, the cost is covered on the Bid Summary form by adding to the material cost a sum equal to 0.20 percent of the direct labor cost. The following items are considered expendable materials:

Augers, ship, Irwin
Batteries, dry cell
Blades, bandsaw, hacksaw
Brooms
Brushes, paint, steel scratch
Burnsmatic
Cans, oil, paint, water
Cloth, drop, canvas, wiping, emery
Cups, drinking
Chamois, skins for cleaning
Chisels, cold
Clamps, hose
Cleaner, hand
Couplings, hose
Dope, pipe
Flux, soldering
Gasoline, white
Hose, water, ends, connections
Oakum
Oil, cutting, jack
Oxygen, tank
Paste, soldering
Plaster, patching
Prestolite, torch sets
Rags, wiping
Rope, hemp
Sash cord
Soap, pulling
Steel wool, welding
Strings, pulling
Tags, pipe
Taps, tape
Thinner, paint
Tubs
Wood lumber

Expediting process

It would be much more difficult to meet completion dates for most projects, and to manage the job with continuity and dispatch, without considerable expediting—and expediting is a process which requires a good set of procedures, as without them it would be extremely frustrating. *Expediting* is defined as the process of ensuring the flow of adequate supplies of material and equipment for continuity of production.

The requirement date for the use of material and equipment is the target which the process of purchasing is destined to achieve, and it is mandatory that it do so. The items that require expediting are material, equipment, tools, and manufacturers' drawings.

The following steps are involved in the expediting process:

- Record the Purchase Order number in the log book by job name.
- File the expedite copy of the Purchase Order in the tickler file by expedite date.
- Expedite on the date indicated according to the following code:
  - XD—Expedite drawings and/or catalog sheets
  - XDD—Expedite more detailed drawings as requested by the contract manager
  - XT—Expedite tubs
  - XR—Expedite recessed fixtures
  - XF—Expedite fittings, poles, brackets, etc.
  - XP—Expedite all parts

  A single Purchase Order may contain most but not all the above codes with expedite dates. It is imperative that the expediting process be accomplished sequentially and complete.
- Contact vendors on indicated expedite dates to determine the following:
  - Status of the acknowledgment to be returned by vendor
  - Vendor shipping date
  - Vendor's capability of meeting the requirement date
- Mail the information request card to out of town vendors. This card is shown in Fig. 6.48.
- Refile the expedite copy in the next sequential date slot if all the information received from supplier does not meet the company requirements. Expedite again, and again, and again until the need is met.
- Refile the expedite copy of the Purchase Order, if all the information is adequate, for checking 1 week before the date for follow-up and confirmation of shipment. Out-of-town vendors are mailed the Request for Proof of Shipment card 2 weeks prior to requirement date.
- Refile the expedite copy in the requirement date pocket.
- Determine whether the material is in the warehouse or on the job. If not in either place, contact the vendor for information on availability.

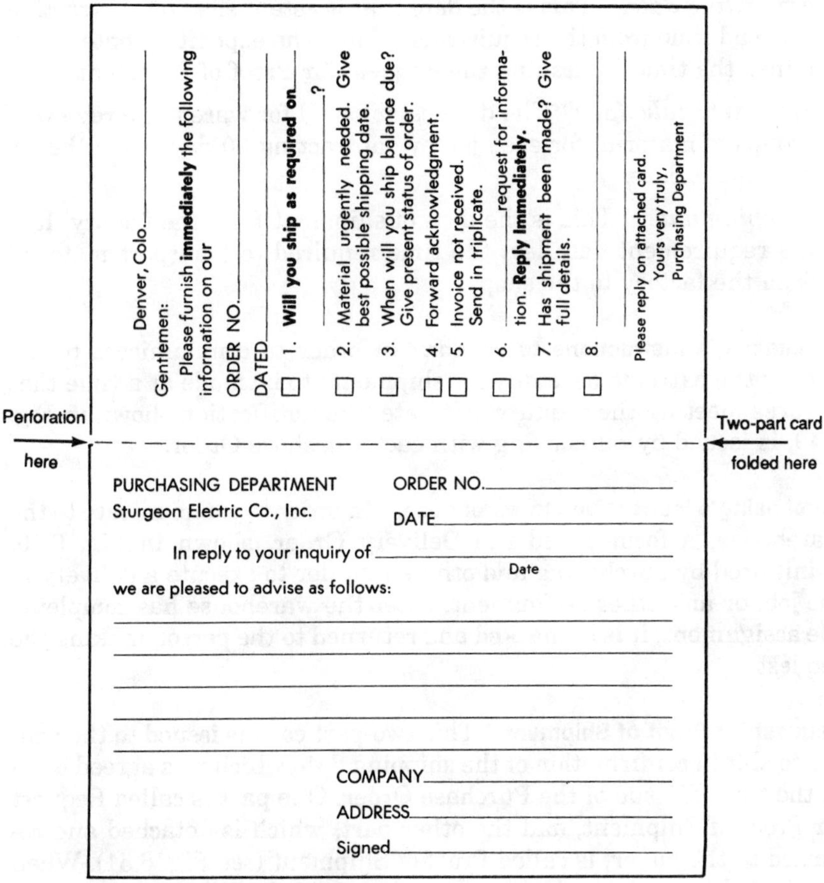

Denver, Colo.______
Gentlemen:
Please furnish **immediately** the following information on our
ORDER NO.______
DATED______
☐ 1. **Will you ship as required on**______?
☐ 2. Material urgently needed. Give best possible shipping date.
☐ 3. When will you ship balance due? Give present status of order.
☐ 4. Forward acknowledgment.
☐ 5. Invoice not received. Send in triplicate.
☐ 6. ______request for information. **Reply Immediately.**
☐ 7. Has shipment been made? Give full details.
☐ 8. ______

Please reply on attached card.
Yours very truly,
Purchasing Department

Perforation here → ← Two-part card folded here

PURCHASING DEPARTMENT
Sturgeon Electric Co., Inc.
ORDER NO.______
DATE______
In reply to your inquiry of ______ (Date)
we are pleased to advise as follows:

COMPANY______
ADDRESS______
Signed______

Figure 6.48 Information request card.

- Pull the expedite copy, discard it, and line out the log record when all the parts have been received.

In the process of expediting it is essential to adhere to uniform terminology. Some definitions follow:

Requirement date: This is the actual time of need on the job.

Schedule date: The day of need as established on the Activity Schedule.

Activity Schedule: The time bar chart prepared by the processor, showing the sequence of activities according to work schedule.

Lead time: Amount of time in weeks which is allowed by the purchasing agent to establish the date of issue for the Purchasing and Expediting Report (See Fig. 6.46).

Expediting date: This is the date that is established by subtracting the lead time from the requirement date. The expediting date determines the time for issuing the Request for Proof of Shipment.

Stock schedule date: The date established for warehouse review of groups of material for each job by subtracting 30 days from the requirement date.

Shipping date: This is the date of shipment from the factory. It is the requirement date less the time required to transport material from the factory to the shop.

Purchasing's instructions to vendor. In order to communicate to the vendor the extreme necessity for shipments to be made at a time that ensures meeting the requirement date, the notification shown in Fig. 6.49, is issued by purchasing with each Purchase Order.

Purchasing's instructions to warehouse. In order to communicate to the warehouse, a form called the Delivery Order, shown in Fig. 6.50, is initiated by purchasing and others in order to execute a delivery to the job, or any other assignment. After the warehouse has completed the assignment, it is so marked and returned to the person making the request.

Request for Proof of Shipment. This two-part card is issued to the vendor to obtain confirmation of the shipping date which was agreed upon at the time of issue of the Purchase Order. One part is called Request for Proof of Shipment, and the other part, which is detached and returned to the buyer, is called Proof of Shipment (see Fig. 6.51). When proof of shipment has been positively established by receipt of the freight bill routing and "pro" number, it is necessary for the expediter to establish the date of actual receipt and notify the contract manager immediately.

Warehouse procedures

It is the responsibility of the warehouse to maintain sufficient stock of standard materials, and to provide for initial job shipments of embedded or rough-in material and Change-Order material. Most initial shipments require truck trailers for job storage as designated by the contract manager, together with job storage facilities and material handling equipment.

It is the objective of the entire material handling procedure to accomplish the following:

- Provide the right material to the job at the designated requirement date

Notification to Vendor:

The attached Purchase Order requires the following paperwork procedures:

1. **Manufacturer's Acknowledgment**

   It is mandatory for us to have immediate acknowledgment of this purchase, showing:

   a. Scheduled shipping date
   b. Vendor's Purchase Order number
   c. Company's Purchase Order number
   d. Job name
   e. Shipping address

2. **Shipping Schedule**

   We have established an intelligent requirement date for delivery to the job. We must have your cooperation in meeting this date. Therefore, your shipping date must allow sufficient time to ensure delivery as specified.

3. **Shop Drawings** (when required)

   These must be available to us as soon as possible, but not later than the date specified, and we must return them promptly. Unnecessary delay of shop drawings might affect shipment.

4. **Material Identification**

   It is necessary for the manufacturer to show Purchase Order identification such as: Type A fixture, or LPB for panel

   Your cooperation with these requirements is necessary to ensure proper service.

Figure 6.49 Notification to Vendor, attached to each Purchase Order.

- Provide proper job facilities for material handling and storage
- Make certain that the tools delivered to the field are in good working order, and the parts are all available

Receiving and Inspection Report

The purpose of the Receiving and Inspection Report, shown in Fig. 6.52, is the following:

- Provide purchasing with evidence of the identity of the contents and its physical condition when shipped to the job by anyone other than

Delivery Order

Date ________ Delivery for ________ Date Required ________

Company Work Order No. ________ Customer Work Order No. ________ Time Required ________

No.	Material Description	Customer P. O. No.	Pick up at	Deliver to

Other Instructions ________

________ By ________

Figure 6.50 Delivery Order.

the contractor's warehouse. Its presence enables purchasing to close the expediting file, clear the invoice for payment, pursue shortages, and correct wrong materials or damaged items.

- Provide a means for the field to check for condition of contents, and to record shipments against Bills of Material to afford better material control.

No Receiving and Inspection Report is required for shipments made to the field from vendor's trucks or common carriers, provided that a packing ticket is available; otherwise a Receiving and Inspection Report should be filled out, listing the items received by catalog number, the shipper, and the Purchase Order number, and including a note on the condition of the items. The details of the shipment must then be recorded on the field copy of the Bill of Material.

Facilities for handling and storing material

Packaging facilities must be provided to accommodate material handling needs for the following operations:

- Putting up material orders in the warehouse

REQUEST FOR **PROOF OF SHIPMENT**

JOBBER__________

JOBBERS P. O. NO__________

STURGEON P. O. NO__________

DESCRIPTION__________

SCHEDULED SHIPPING DATE__________

PLEASE FORWARD THE ATTACHED POST CARD
WITH COMPLETE INFORMATION ON:

PROOF OF SHIPMENT

This is a vital part of your service to us. Please be prompt.

Perforation here — Two-part card folded here

PROOF OF SHIPMENT

DESCRIPTION__________

JOBBER P. O. NO.__________

STURGEON P. O. NO.__________

SHIPPING DATE__________

ROUTE__________

"PRO" NO.__________

Mfg.__________ By__________

Figure 6.51 Request for Proof of Shipment card.

- Transporting material to the warehouse
- Receiving and storing material on the job
- Carting material to the point of usage on the job
- Conveying material back to the shop for subsequent credit

The equipment and containers must meet the following requirements:

- Available in uniform sizes and colors
- Partially open-ended for visual observation of contents
- Stackable
- May be fitted to a rack and easily detached
- Made of sturdy material
- Made of lightweight material

Receiving and Inspection Report

Receiving Tally No. 4065

ITEM NO.	QUANTITY	CATALOG NO.	DESCRIPTION OF ARTICLE
2			
3			
4			
5			
22			
23			
24			

P. O. No. _____

Vendor _____ Date Rec. _____

Shipper _____ Shipped From _____

Via _____ Waybill No. _____ Unloaded At _____

Remarks _____

Received By _____ Checked By _____ Approved By _____

Matched With P. O. By _____

Matched With Invoice By _____ Invoice No. _____ Date _____ Amount Approved _____

Figure 6.52 Receiving and Inspection Report.

The following equipment is necessary for material handling in both the shop warehouse and on the job:

- Plastic bins
- A container rack for storing bins
- A three-tiered cart on wheels
- A heavy-duty four-wheel industrial truck
- A two-wheeled trunk carrier–type truck
- Metal tote boxes to fit on the three-tiered cart
- A tool gang box
- A material gang box for cartons

Material handling equipment. Figure 6.53 illustrates the manner in which material handling equipment is coordinated to accommodate conditions in the warehouse as well as on the job. This equipment is

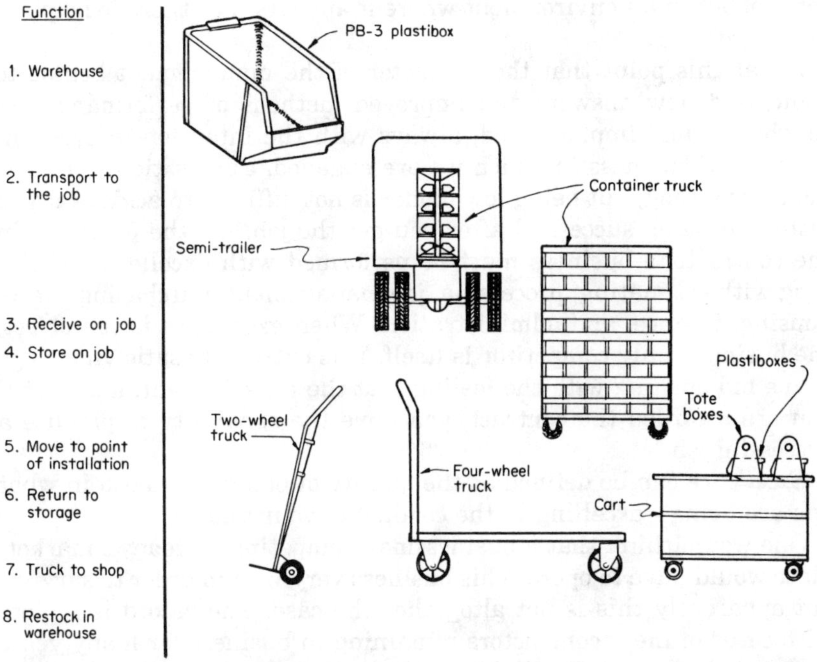

Figure 6.53 Material handling devices and equipment.

used for stocking material, loading it in a trailer, and conveniently handling it when unloading on a job.

In the final analysis, the contractor's responsibility is to provide the right material for the job, in the right quantity, and at the right time. When this responsibility is properly handled, the contractor will be more competitive, more productive, more efficient and more successful. Fortunately, the system which makes it all happen is readily adaptable to computer application; *unfortunately,* it makes mandatory the ongoing training and indoctrination of the entire organization.

Summary: The Goal of Excellence in Operations

Like all other pursuits, the business of electrical contracting requires ongoing motivation to fulfill its economic responsibilities, which change from time to time as the population increases and competition becomes more intense. As competitive conditions intensify, the need for better performance on the part of a business increases to the point where changes in operations become mandatory. How can changes come about in an environment where it appears that there is no place else to go?

It's at this point that the character of the management begins to show, and new answers and improved methods of performance are sought out and implemented, always with the intention to excel. In the type of business in which we are engaged, competition to be low bidder is strong, but being low bidder is not difficult to achieve if you disregard being successful after you get the job. All the functions in the contractor's business must be performed with excellence, beginning with estimating, processing, job management, purchasing, warehousing, finance, and administration. When excellence is the object, the business' only competitor is itself. It is extremely satisfying to go into a bid opening with the feeling that the price is right, and that if you are awarded the contract, you have the capability to produce a successful job.

Excellence can be defined as the quality of being very good in what you are doing—excelling in the conduct of your affairs.

One would think that a businessman competing on today's marketplace would have to operate his business very well in order to survive, but apparently this is not altogether the case. The record is replete with cases of inept contractors remaining in business for many years before their demise. Topflight journeymen can make up for the inadequacies of the contractor, but it's hazardous to rely on them completely.

One of the fortunate things about our industry is that it is composed of competent people. *Competence* can be defined as "having requisite ability, sufficient for the performance of the tasks to which one is assigned." In a free society such as ours, competence is an elementary duty, but excellence implies more than competence—it implies a striving for the highest standards of performance. One of the *unfortunate* things about our business is that there are so many different ways of accomplishing the same thing. Nevertheless there has to be a "best" way of doing things which results in less time being required, a lower cost of operation, the most acceptable method from a mechanical standpoint, and the most generally acceptable from a technical standpoint. The most adroit persons do things the best way unconsciously, and this is the essence of excellence in operations.

But we're not talking about doing things unconsciously, we're talking about doing things according to a business system where all the details are spelled out, where employees are trained to fit into the system, and where each operation is related to the entire system. The efficient installation of an electrical construction job and the achievement of productive man-hours of work are difficult to accomplish. There are many factors involved, some of which are controllable by the electrical contractor and some not. The planning, scheduling, and mastery of the details of the installation by everyone concerned are mandatory if satisfactory results are to be achieved, and this demands close coordination between all functions of the business and the field personnel who make the installation.

In order for the business to achieve the level of performance necessary to provide outstanding service, it must be staffed by individuals who have been properly trained for the tasks to which they have been assigned. These people must also be provided with suitable forms of communication, so that one specialist can pick up where another leaves off for the achievement of maximum productivity. *Productivity* may be defined as "the capability of effecting or contributing to effective production." *Capability* relates to the quality of having general efficiency, with physical and mental ability. It should be recognized that a person could have physical and mental ability without having general efficiency. *Efficiency* is defined as "effective operation as measured by a comparison of production with cost." From this it appears that the whole matter of productivity is to be able to bring about effective production within the realm of competitive cost limitations."

The goal of excellence in operations is within the grasp of any person, company, or organization that is willing to aspire to a performance of duty that exceeds ordinary expectations, and the results will provide the degree of productivity required for successful operations.

Appendix

Box, Concrete, with Cover

Date 1984

Universal no	Description	Type of construction	Man-hours per each
54521	Box, concrete with back plate and stud	Flat slab, on wood	0.40
54521	Box, concrete with back plate and stud	Flat slab, on steel	0.55
54521	Box, concrete with back plate and stud	Steel pan outlets in beams	0.50
54521	Box, concrete with back plate and stud	Flat slab, tile filler	0.45

Factoring

Conditions of Usage

Fastening box to form included
Conduit terminals not included
Masking tape on back and bottom edges included
Back plate with stud included

Conditions of Installation

Work done by proper number of average journeymen using proper tools
Work installed in a local area, at ground level, and at a height of not over 14 ft
Work done with normal concentrations of material, with normal material delivery, and at normal rate of progress
Work done with normal access to job, no impediments
Work performed at normal working temperatures

Distribution of Time, %

Study Plans	Order Material	Receive and Store Material	Move to Point of Use	Tool Up	Measure	Install	Nonproductive	Total
8	2	2	2	6	10	60	10	100

Labor Unit Table 1.

Box, Outlet, 4-Inch Octagonal

Universal no	Description	Type of construction	Man-hours per each
54151	Box, 4-inch octagonal	Exposed	0.16
54151	Box, 4-inch octagonal	Furred	0.19
54151	Box, 4-inch octagonal	Masonry flush	0.45
54151	Box, 4-inch octagonal	Reinforced concrete	0.32
54151	Box, 4-inch octagonal	Metal frame	0.21
54151	Box, 4-inch octagonal	Wood frame	0.21
54151	Hanger, box		0.20
54151	Cover, box for plaster or blank		0.12
54151	Stud, fixture		0.12

Factoring

Units based on a normal quantity of 24 for installation density
For quantity of 72 in normal working area reduce unit 14%
For quantity of 144 in normal working area reduce unit 17%

Conditions of Usage

Fastenings not included
Conduit terminals not included
One-man installation

Conditions of Installation

Work done by proper number of average journeymen using proper tools
Work installed in a local area, at ground level, and at a height of not over 14 ft
Work done with normal concentrations of material, with normal material delivery, and at normal rate of progress
Work done with normal access to job, no impediments
Work performed at normal working temperatures

Distribution of Time, %

Study Plans	Order Material	Receive and Store Material	Move to Point of Use	Tool Up	Measure	Install	Nonproductive	Total
8	2	3	5	5	10	57	10	100

Labor Unit Table 2.

Box,Outlet, 4-Inch Square

Universal no.	Description	Type of construction	Man-hours per each
52151	Box,4-inch square	Metal frame	0.22
52151	Box,4-inch square	Wood frame	0.22
52151	Box,4-inch square	Masonry flush	0.45
52151	Box,4-inch square	Exposed	0.16
52151	Box,4-inch square	Reinforced concrete	0.32
52151	Hanger,box - expanded type		0.20
52151	Cover,box - plaster ring,outlet,or blank		0.12
52151	Stud,fixture		0.12

Factoring

Units based on a normal quantity of 24 for installation density
For quantity of 72 in normal working area reduce unit 14%
For quantity of 144 in normal working area reduce unit 17%

Conditions of Usage

Fastenings not included
Conduit terminals not included
One-man installation

Conditions of Installation

Work done by proper number of average journeymen using proper tools
Work installed in a local area, at ground level, and at a height of not over 14 ft
Work done with normal concentrations of material, with normal material delivery, and at normal rate of progress
Work done with normal access to job, no impediments
Work performed at normal working temperatures

Distribution of Time, %

Study Plans	Order Material	Receive and Store Material	Move to Point of Use	Tool Up	Measure	Install	Nonproductive	Total
8	2	3	5	5	10	57	10	100

Labor Unit Table 3.

Box, Pull			
Size, in	Weight, lb	Description	Man-hours per each
14-gauge			
12x12x4	12	One-man	1.20
12x18x4	16	One-man	1.70
12x24x4	21	One-man	2.00
18x18x6	23	One-man	2.20
18x24x6	32	One-man	2.50
18x30x6	49	One-man	2.90
24x24x6	42	One-man	3.00
24x30x8	54	One-man	3.20
12-gauge			
24x36x8	88	Two-man	7.15
30x30x8	90	Two-man	7.30
30x36x8	105	Two-man	7.80
36x36x12	140	Two-man	8.50
36x42x12	204	Two-man	9.60
42x48x12	256	Two-man	11.00
48x48x16	317	Two-man	13.50
48x60x16	348	Two-man	15.00

Factoring

For installation at height of 15 ft to 25 ft multiply unit by 1.50
For installation at height of 26 to 35 ft multiply unit by 1.70

Conditions of Usage

All 10-gauge and 12-gauge boxes require 2 men for installation
Conduit terminals or knockouts not included
Fastenings not included
Brackets, shelves, or hangers not included

Conditions of Installation

Work done by proper number of average journeymen using proper tools
Work installed in a local area, at ground level, and at a height of not over 14 ft
Work done with normal concentrations of material, with normal material delivery, and at normal rate of progress
Work done with normal access to job, no impediments
Work performed at normal working temperatures

Distribution of Time, %

Study Plans	Order Material	Receive and Store Material	Move to Point of Use	Tool Up	Measure	Install	Nonproductive	Total
5	2	3	5	5	5	65	10	100

Labor Unit Table 4.

Busway, Feeder, Aluminum, Encapsulated		
Size, A	Description	Man-hours per ft
800	3-Pole busway,feeder,aluminum encapsulated	0.135
1000	3-Pole busway,feeder,aluminum encapsulated	0.148
1350	3-Pole busway,feeder,aluminum encapsulated	0.160
1600	3-Pole busway,feeder,aluminum encapsulated	0.176
2000	3-Pole busway,feeder,aluminum encapsulated	0.189
2500	3-Pole busway,feeder,aluminum encapsulated	0.210
3000	3-Pole busway,feeder,aluminum encapsulated	0.250
4000	3-Pole busway,feeder,aluminum encapsulated	0.325
800	4-PFN busway,feeder,aluminum encapsulated	0.150
1000	4-PFN busway,feeder,aluminum encapsulated	0.175
1350	4-PFN busway,feeder,aluminum encapsulated	0.190
1600	4-PFN busway,feeder,aluminum encapsulated	0.205
2000	4-PFN busway,feeder,aluminum encapsulated	0.220
2500	4-PFN busway,feeder,aluminum encapsulated	0.280
3000	4-PFN busway,feeder,aluminum encapsulated	0.320
4000	4-PFN busway,feeder,aluminum encapsulated	0.350

Factoring	For parallel runs reduce units 17% Factor for risers according to ceiling height and number of floors
Conditions of Usage	Fastenings and/or hangers not included Fork lift operator included Cost of fork lift or hoist not included Increase units 50% if fork lift or hoist is not available
Conditions of Installation	Work done by proper number of average journeymen using proper tools Work installed in a local area, at ground level, and at a height of not over 14 ft Work done with normal concentrations of material, with normal material delivery, and at normal rate of progress Work done with normal access to job, no impediments Work performed at normal working temperatures

Distribution of Time, %	Study Plans	Order Material	Receive and Store Material	Move to Point of Use	Tool Up	Measure	Install	Nonproductive	Total
	3	2	3	3	25	5	49	10	100

Labor Unit Table 5.

Busway, Feeder, Aluminum, Encapsulated Fittings							
Size, A	Poles	Man-hours per each					
		Terminal	Ell	Tee	Center tap box	End cap fitting	Hanger
800	3	2.85	0.40	0.55	2.50	0.50	0.30
1000	3	2.85	0.45	0.60	2.50	0.50	0.30
1350	3	3.00	0.50	0.70	2.50	0.50	0.30
1600	3	3.50	0.90	1.10	2.50	0.50	0.30
2000	3	3.50	1.00	1.20	3.50	0.75	1.30
2500	3	4.00	1.20	1.30	3.50	0.75	1.30
3000	3	4.50	1.80	2.70	4.00	0.75	1.30
4000	3	6.00	2.00	2.90	4.00	0.75	1.30
800	4	3.30	0.40	0.60	2.50	0.60	0.30
1000	4	3.30	0.45	0.65	2.50	0.60	0.30
1350	4	3.60	0.50	0.80	2.50	0.60	0.30
1600	4	3.70	0.90	1.20	2.50	0.60	0.30
2000	4	4.00	1.00	1.30	3.50	0.90	1.30
2500	4	5.00	1.20	1.45	3.50	0.90	1.30
3000	4	5.40	2.00	2.85	4.00	0.90	1.30
4000	4	7.10	2.30	3.15	4.00	0.90	1.30

Factoring

Conditions of Usage

Fastening not included for hanger

Conditions of Installation

Work done by proper number of average journeymen using proper tools
Work installed in a local area, at ground level, and at a height of not over 14 ft
Work done with normal concentrations of material, with normal material delivery, and at normal rate of progress
Work done with normal access to job, no impediments
Work performed at normal working temperatures

Distribution of Time, %

Study Plans	Order Material	Receive and Store Material	Move to Point of Use	Tool Up	Measure	Install	Nonproductive	Total
3	2	3	3	25	5	49	10	100

Labor Unit Table 6.

Cable, Communications		
Size, no	Description	Man-hours per ft
2-22	Cable,plastic/plastic jacket	0.014
4-22	Cable,plastic/plastic jacket	0.016
6-22	Cable,plastic/plastic jacket	0.020
8-22	Cable,plastic/plastic jacket	0.024
10-22	Cable,plastic/plastic jacket	0.026
12-22	Cable,plastic/plastic jacket	0.030
2-16	Cable,plastic/plastic jacket	0.020
3-16	Cable,plastic/plastic jacket	0.025
4-16	Cable,plastic/plastic jacket	0.030
5-16	Cable,plastic/plastic jacket	0.040
7-16	Cable,plastic/plastic jacket	0.045
9-16	Cable,plastic/plastic jacket	0.050
12-16	Cable,plastic/plastic jacket	0.060

Factoring	
Conditions of Usage	Cable pulled in raceway or run concealed Terminals not included Fastenings not included Cable furnished on reels
Conditions of Installation	Work done by proper number of average journeymen using proper tools Work installed in a local area, at ground level, and at a height of not over 14 ft Work done with normal concentrations of material, with normal material delivery, and at normal rate of progress Work done with normal access to job, no impediments Work performed at normal working temperatures

Distribution of Time, %	Study Plans	Order Material	Receive and Store Material	Move to Point of Use	Tool Up	Measure	Install	Nonproductive	Total
	5	2	2	2	5	5	69	10	100

Labor Unit Table 7.

Cable, Nonmetallic, Sheathed

Date 1984

Size, no	Description	Man-hours per ft	Man-hours per each
12	Cable,NMC 12-2/grounded	.035	
12	Cable,NMC 12-3/grounded	.042	
10	Cable,NMC 10-2/grounded	.050	
10	Cable,NMC 10-3/grounded	.055	
8	Cable,NMC 8-1	.030	
	Connectors,cable		.08
	Straps and staples		.03

Factoring

Conditions of Usage

Unit for connector includes terminal
Material furnished in coils

Conditions of Installation

Work done by proper number of average journeymen using proper tools
Work installed in a local area, at ground level, and at a height of not over 14 ft
Work done with normal concentrations of material, with normal material delivery, and at normal rate of progress
Work done with normal access to job, no impediments
Work performed at normal working temperatures

Distribution of Time, %

Study Plans	Order Material	Receive and Store Material	Move to Point of Use	Tool Up	Measure	Install	Nonproductive	Total
5	2	2	5	4	4	68	10	100

Labor Unit Table 8.

Cabinets, Telephone

Size, in	Description	Man-hours per each
30x12	Cabinet,14 gauge surface	2.90
30x18	Cabinet,14 gauge surface	3.40
30x24	Cabinet,14 gauge surface	3.90
30x30	Cabinet,14 gauge surface	4.50
36x30	Cabinet,14 gauge surface	5.00
48x24	Cabinet,12 gauge surface	7.45
48x30	Cabinet,12 gauge surface	8.20
48x36	Cabinet,12 gauge surface	9.00
60x60	Cabinet,10 gauge surface	18.00

Factoring

Conditions of Usage

All cabinets 4 in deep. Plywood back board included.
Trim with door included
Fastenings not included
Conduit entrances not included
For flush mounting add 10%

Conditions of Installation

Work done by proper number of average journeymen using proper tools
Work installed in a local area, at ground level, and at a height of not over 14 ft
Work done with normal concentrations of material, with normal material delivery, and at normal rate of progress
Work done with normal access to job, no impediments
Work performed at normal working temperatures

Distribution of Time, %

Study Plans	Order Material	Receive and Store Material	Move to Point of Use	Tool Up	Measure	Install	Nonproductive	Total
4	2	3	5	5	5	66	10	100

Labor Unit Table 9.

Conduit, Aluminum, Rigid

Size, in	Description	Weight, lb	Man-hours per ft
1/2	Conduit,aluminum rigid one-man	28.1	.018
3/4	Conduit,aluminum rigid one-man	37.4	.022
1	Conduit,aluminum rigid one-man	54.5	.033
1-1/4	Conduit,aluminum rigid one-man	71.6	.042
1-1/2	Conduit,aluminum rigid one-man	88.7	.052
2	Conduit,aluminum rigid one-man	118.5	.071
2-1/2	Conduit,aluminum rigid one-man	187.5	0.090
3	Conduit,aluminum rigid one-man	246.3	0.120
3-1/2	Conduit,aluminum rigid one-man	295.6	0.145
4	Conduit,aluminum rigid one-man	350.2	0.174
5	Conduit,aluminum rigid two-man	478.9	0.400
6	Conduit,aluminum rigid two-man	630.4	0.580

Note:1.Weight shown in pounds per hundred feet.
2.Steel fittings and ells are used with aluminum rigid conduit.

Factoring

Increase units for different types of construction
Increase units for height factor
Increase units for risers (use height factor and number of floors factor)
Reduce units for parallel runs

Conditions of Usage

Bends and offsets not included. Fastening labor not included
Hangers not included
Terminals not included

Conditions of Installation

Work done by proper number of average journeymen using proper tools
Work installed in a local area, at ground level, and at a height of not over 14 ft
Work done with normal concentrations of material, with normal material delivery, and at normal rate of progress
Work done with normal access to job, no impediments
Work performed at normal working temperatures

Distribution of Time, %

Study Plans	Order Material	Receive and Store Material	Move to Point of Use	Tool Up	Measure	Install	Nonproductive	Total
4	2	4	4	10	4	62	10	100

Labor Unit Table 10.

Conduit, EMT Branch Circuit

Type of construction	Access factor, %	Bends per 100 ft	Offsets per 100 ft	Man-hours per ft 1/2 in	3/4 in	1 i.
Deck conduit turned up	15	5	5	0.031	0.041	0.054
Deck, flat	15	4	4	0.027	0.035	0.048
Exposed on surface		4	4	0.025	0.030	0.044
Furred ceiling	10	2	2	0.023	0.030	0.042
Masonry walls	15	3	3	0.027	0.035	0.047
Metal stud walls		3	3	0.023	0.030	0.041
Notched joists	10	4	6	0.030	0.039	0.053
Reinforced concrete walls	5	3	3	0.025	0.032	0.043
Wood frame	10	4	4	0.028	0.035	0.049

Factoring

Conditions of Usage

One man used to make the installation — 1/2 in; 3/4 in; 1 in
Designated number of bends, offsets, access factor included
Fastening and terminals not included
Labor for installing couplings and connectors not included

Conditions of Installation

Work done by proper number of average journeymen using proper tools
Work installed in a local area, at ground level, and at a height of not over 14 ft
Work done with normal concentrations of material, with normal material delivery, and at normal rate of progress
Work done with normal access to job, no impediments
Work performed at normal working temperatures

Distribution of Time, %

Study Plans	Order Material	Receive and Store Material	Move to Point of Use	Tool Up	Measure	Install	Nonproductive	Total
4	2	4	4	8	4	64	10	100

Labor Unit Table 11.

Conduit, EMT, Steel			
Size, in	Weight, lb per 10-ft length	Description	Man-hours per ft
1/2	3.0	One-man installation	0.016
3/4	4.5	One-man installation	0.022
1	6.9	One-man installation	0.032
1-1/4	10.4	One-man installation	0.050
1-1/2	11.9	One-man installation	0.060
2	14.4	One-man installation	0.069
2-1/2	23.0	One-man installation	0.108
3	27.0	One-man installation	0.130
3-1/2	36.5	One-man installation	0.168
4	40.0	One-man installation	0.190

Factoring

Increase units for height factor: Increase units for risers
Increase 1/2" - 3/4" units 50% for branch circuit work
Reduce units for parallel runs: 2 - 0.85 multiplier
3 - 0.80 multiplier
4 - 0.77 multiplier
5 - 0.74 multiplier

Conditions of Usage

Coupling labor not included
Hangers or fastenings not included
Connectors not included
Handling based on maximum size bundles

Conditions of Installation

Work done by proper number of average journeymen using proper tools
Work installed in a local area, at ground level, and at a height of not over 14 ft
Work done with normal concentrations of material, with normal material delivery, and at normal rate of progress
Work done with normal access to job, no impediments
Work performed at normal working temperatures

Distribution of Time, %

Study Plans	Order Material	Receive and Store Material	Move to Point of Use	Tool Up	Measure	Install	Nonproductive	Total
7	2	5	4	8	6	58	10	100

Labor Unit Table 12.

Conduit, GRS

Date 1984

Size, in	Weight, lb per 10-ft length	Description	Man-hours per ft
1/2	7.9	One-man operation	0.035
3/4	10.5	One-man operation	0.042
1	15.3	One-man operation	0.060
1-1/4	20.1	One-man operation	0.076
1-1/2	24.9	One-man operation	0.090
2	33.2	One-man operation	0.120
2-1/2	52.7	Two-man operation	0.320
3	68.2	Two-man operation	0.370
3-1/2	83.1	Two-man operation	0.418
4	97.2	Two-man operation	0.467
5	131.3	Two-man operation	0.580
6	174.5	Two-man operation	0.730

Factoring

Increase units for different types of construction
Increase units for height factor
Increase units for risers (use height factor and number of floors factor)
Reduce units for parallel runs

Conditions of Usage

Fastening labor not included
Hangers not included
Terminals not included
1 cut per thread per 200 ft included
1 bend per 200 ft included

Conditions of Installation

Work done by proper number of average journeymen using proper tools
Work installed in a local area, at ground level, and at a height of not over 14 ft
Work done with normal concentrations of material, with normal material delivery, and at normal rate of progress
Work done with normal access to job, no impediments
Work performed at normal working temperatures

Distribution of Time, %

Study Plans	Order Material	Receive and Store Material	Move to Point of Use	Tool Up	Measure	Install	Nonproductive	Total
4	2	4	4	10	4	62	10	100

Labor Unit Table 13.

Connectors, Wire, Compression Type		
Wire size no.	Man-hours per each	
8	0.25	
6	0.35	
4	0.40	
2	0.48	
1	0.51	
1/0	0.58	
2/0	0.63	
3/0	0.72	
4/0	0.82	
250	0.92	
300	0.99	
350	1.07	
400	1.14	
500	1.28	
		Note: Wire size identifies connector size.

Factoring	
Conditions of Usage	Cutting and skinning wire included Compression tool without dies used for indentation Taping or cover included
Conditions of Installation	Work done by proper number of average journeymen using proper tools Work installed in a local area, at ground level, and at a height of not over 14 ft Work done with normal concentrations of material, with normal material delivery, and at normal rate of progress Work done with normal access to job, no impediments Work performed at normal working temperatures

Distribution of Time, %	Study Plans	Order Material	Receive and Store Material	Move to Point of Use	Tool Up	Measure	Install	Nonproductive	Total
	2	2	2	4	8	6	66	10	100

Labor Unit Table 14.

Devices and Plates

Size	Description	Man-hours per each
15 A	Receptacles, flush grounded type	0.15
20 A	Receptacles, flush grounded type	0.23
20 A	Receptacles, flush polarized	0.40
50 A	Receptacles, flush polarized	0.47
10 A	Switches, flush SP	0.15
10 A	Switches, flush 3-way	0.23
10 A	Switches, flush 4-way	0.40
1 Gang	Plates, flush device	0.12
2 Gang	Plates, flush device	0.17
3 Gang	Plates, flush device	0.25
2-Wire	Caps, attachment grounded	0.30
3-Wire	Caps, attachment grounded	0.40

Factoring

Conditions of Usage

Devices installed with branch circuit wire
Plates installed after painting walls

Conditions of Installation

Work done by proper number of average journeymen using proper tools
Work installed in a local area, at ground level, and at a height of not over 14 ft
Work done with normal concentrations of material, with normal material delivery, and at normal rate of progress
Work done with normal access to job, no impediments
Work performed at normal working temperatures

Distribution of Time, %

Study Plans	Order Material	Receive and Store Material	Move to Point of Use	Tool Up	Measure	Install	Nonproductive	Total
10	5	2	2	2		69	10	100

Labor Unit Table 15.

Ells, Conduit Rigid

Size, in	Man-hours per each		
	Ells	Bends	Offsets
1/2		0.28	0.45
3/4		0.38	0.55
1	0.62	0.52	0.75
1-1/4	0.82	0.55	0.90
1-1/2	0.97	0.72	1.10
2	1.29	1.00	1.38
2-1/2	2.48	2.20	2.50
3	2.98	2.70	3.00
3-1/2	3.48	3.20	3.55
4	3.94	3.65	4.00
5	4.95	4.50	5.50

Factoring

Conditions of Usage

Pipe dope application included where required
Field bends and offsets made with mechanical tools
Type of conduit ells includes 45°, 90°, and short ells
Fastenings not included

Conditions of Installation

Work done by proper number of average journeymen using proper tools
Work installed in a local area, at ground level, and at a height of not over 14 ft
Work done with normal concentrations of material, with normal material delivery, and at normal rate of progress
Work done with normal access to job, no impediments
Work performed at normal working temperatures

Distribution of Time, %

Study Plans	Order Material	Receive and Store Material	Move to Point of Use	Tool Up	Measure	Install	Nonproductive	Total
Ells- 2	2	2	2	18	12	52	10	100
Bends-2	0	0	0	18	12	58	10	100

Labor Unit Table 16.

Ells, EMT Steel			
Size, in	Man-hours per each		
	Regular	Field bend	Offset
1/2		0.12	0.16
3/4		0.15	0.20
1	0.16	0.20	0.30
1-1/4	0.23	0.25	0.36
1-1/2	0.25	0.28	0.40
2	0.32	0.35	0.50
2-1/2	0.35	0.40	0.60
3	0.45	0.50	0.70
3-1/2	0.50	0.60	0.80
4	0.60	0.70	0.90

Factoring	Factor for height of building
Conditions of Usage	Coupling labor not included
Conditions of Installation	Work done by proper number of average journeymen using proper tools Work installed in a local area, at ground level, and at a height of not over 14 ft Work done with normal concentrations of material, with normal material delivery, and at normal rate of progress Work done with normal access to job, no impediments Work performed at normal working temperatures

Distribution of Time, %	Study Plans	Order Material	Receive and Store Material	Move to Point of Use	Tool Up	Measure	Install	Nonproductive	Total
	2	2	2	2	20	15	47	10	100

Labor Unit Table 17.

Fastenings, Anchor, Cinch

Size, in	Man-hours per each
8/32	0.15
10/24	0.17
12/24	0.22
1/4	0.24
5/16	0.26
3/8	0.30
1/2	0.35
5/8	0.43
3/4	0.48

Factoring

Conditions of Usage

Units based on using electric drill with carbide bits
Anchors set in 1/4 in from surface
Masonry includes cinder block, brick, tile, stone

Conditions of Installation

Work done by proper number of average journeymen using proper tools
Work installed in a local area, at ground level, and at a height of not over 14 ft
Work done with normal concentrations of material, with normal material delivery, and at normal rate of progress
Work done with normal access to job, no impediments
Work performed at normal working temperatures

Distribution of Time, %

Study Plans	Order Material	Receive and Store Material	Move to Point of Use	Tool Up	Measure	Install	Nonproductive	Total
2	2	2	2	16	10	56	10	100

Labor Unit Table 18.

Fastening, Anchor, Plastic or Fiber

Size, no	Man-hours per each
6	0.07
8	0.10
10	0.12
12-3/4	0.14
12-1	0.16
12-1-1/4	0.18
14-1	0.18
14-1-1/2	0.20
14-2	0.22

Factoring

Conditions of Usage

Units based on using electric drill with carbide bits
Anchors set in 1/4 in from surface
Masonry includes cinder block, brick, tile, stone

Conditions of Installation

Work done by proper number of average journeymen using proper tools
Work installed in a local area, at ground level, and at a height of not over 14 ft
Work done with normal concentrations of material, with normal material delivery, and at normal rate of progress
Work done with normal access to job, no impediments
Work performed at normal working temperatures

Distribution of Time, %

Study Plans	Order Material	Receive and Store Material	Move to Point of Use	Tool Up	Measure	Install	Nonproductive	Total
2	2	2	2	16	10	56	10	100

Labor Unit Table 19.

Fastenings, Lag Bolt		
Size, in	**Description**	**Man-hours per each**
1/4	Bolts,lag,2 to 3 in	0.08
5/16	Bolts,lag,2 to 4 in	0.11
3/8	Bolts,lag,2-1/2 to 6 in	0.16
1/2	Bolts,lag,3 to 6 in	0.20
3/4	Bolts,lag,5 to 6 in	0.24

Factoring

Conditions of Usage

Conditions of Installation

Work done by proper number of average journeymen using proper tools
Work installed in a local area, at ground level, and at a height of not over 14 ft
Work done with normal concentrations of material, with normal material delivery, and at normal rate of progress
Work done with normal access to job, no impediments
Work performed at normal working temperatures

Distribution of Time, %

Study Plans	Order Material	Receive and Store Material	Move to Point of Use	Tool Up	Measure	Install	Nonproductive	Total
2	2	2	2	16	10	56	10	100

Labor Unit Table 20.

Fastenings, Machine Bolt			
Size, in	Man-hours per each		
	Tapped in steel	Wood	Steel
3/16	0.30	0.15	0.18
1/4	0.33	0.20	0.22
5/16	0.35	0.25	0.30
3/8	0.40	0.32	0.35
7/16	0.43	0.35	0.40
1/2	0.52	0.40	0.45

Factoring	
Conditions of Usage	
Conditions of Installation	Work done by proper number of average journeymen using proper tools Work installed in a local area, at ground level, and at a height of not over 14 ft Work done with normal concentrations of material, with normal material delivery, and at normal rate of progress Work done with normal access to job, no impediments Work performed at normal working temperatures

Distribution of Time, %	Study Plans	Order Material	Receive and Store Material	Move to Point of Use	Tool Up	Measure	Install	Nonproductive	Total
	2	2	2	2	16	10	56	10	100

Labor Unit Table 21.

Fastenings, Toggle Bolt

Size, in	Man-hours per each	
1/8	0.12	All Bolts 3" Long
3/16	0.17	
1/4	0.22	
5/16	0.31	
3/8	0.35	

Factoring	Cinder Block Construction
Conditions of Usage	Units based on using electrical drill with carbide bits
Conditions of Installation	Work done by proper number of average journeymen using proper tools Work installed in a local area, at ground level, and at a height of not over 14 ft Work done with normal concentrations of material, with normal material delivery, and at normal rate of progress Work done with normal access to job, no impediments Work performed at normal working temperatures

Distribution of Time, %	Study Plans	Order Material	Receive and Store Material	Move to Point of Use	Tool Up	Measure	Install	Nonproductive	Total
	2	2	2	2	16	10	56	10	100

Labor Unit Table 22.

Fittings, Conduit, EMT, Steel

Size, in	Description	Man-hours per each: Free air	Exposed	Masonry
1/2	Connectors,EMT	0.11	0.12	0.22
3/4	Connectors,EMT	0.13	0.14	0.24
1	Connectors,EMT	0.15	0.16	0.26
1-1/4	Connectors,EMT	0.22	0.28	
1-1/2	Connectors,EMT	0.24	0.30	
2	Connectors,EMT	0.30	0.36	
2-1/2	Connectors,EMT	0.35	0.45	
3	Connectors,EMT	0.40	0.50	
3-1/2	Connectors,EMT	0.50	0.60	
4	Connectors,EMT	0.55	0.70	
1/2	Couplings,EMT	0.06		
3/4	Couplings,EMT	0.07		
1	Couplings,EMT	0.09		
1-1/4	Couplings,EMT	0.11		
1-1/2	Couplings,EMT	0.13		
2	Couplings,EMT	0.15		
2-1/2	Couplings,EMT	0.18		
3	Couplings,EMT	0.20		
3-1/2	Couplings,EMT	0.30		
4	Couplings,EMT	0.35		
All sizes	Straps and clamps	0.06		

Factoring

Conditions of Usage

Connector terminals do not include cutting K.O.
Coupling and clamp labor for handling and placing
Each connector represents conduit terminal

Conditions of Installation

Work done by proper number of average journeymen using proper tools
Work installed in a local area, at ground level, and at a height of not over 14 ft
Work done with normal concentrations of material, with normal material delivery, and at normal rate of progress
Work done with normal access to job, no impediments
Work performed at normal working temperatures

Distribution of Time, %

Study Plans	Order Material	Receive and Store Material	Move to Point of Use	Tool Up	Measure	Install	Nonproductive	Total
2	2	2	2	15	15	52	10	100

Labor Unit Table 23.

Fittings, Entrance, Threaded	
Size, in	Man-hours per each
1/2	0.35
3/4	0.45
1	0.62
1-1/4	0.82
1-1/2	0.97
2	1.29
2-1/2	1.98
3	2.58
3-1/2	3.09
4	3.64
5	4.95

Factoring	
Conditions of Usage	Covers included Hub terminal included
Conditions of Installation	Work done by proper number of average journeymen using proper tools Work installed in a local area, at ground level, and at a height of not over 14 ft Work done with normal concentrations of material, with normal material delivery, and at normal rate of progress Work done with normal access to job, no impediments Work performed at normal working temperatures

Distribution of Time, %	Study Plans	Order Material	Receive and Store Material	Move to Point of Use	Tool Up	Measure	Install	Nonproductive	Total
	2	2	2	2	18	4	60	10	100

Labor Unit Table 24.

Fittings, Conduit, Rigid, Hub Type			
Size, in	Man-hours per each		
	One hub	Two hub	Three hub
1/2	0.25	0.35	0.45
3/4	0.32	0.45	0.58
1	0.45	0.62	0.79
1-1/4	0.60	0.82	1.04
1-1/2	0.72	0.97	1.24
2	0.96	1.29	1.62
2-1/2	1.48	1.98	2.48
3	1.94	2.58	3.22
3-1/2	2.32	3.09	3.86
4	2.74	3.64	4.54
5	3.75	4.95	6.15

Factoring	
Conditions of Usage	Cover installation included One full hub terminal, and one or two 3/4-in hubs as applicable are included Gaskets not included
Conditions of Installations	Work done by proper number of average journeymen using proper tools Work installed in a local area, at ground level, and at a height of not over 14 ft Work done with normal concentrations of material, with normal material delivery, and at normal rate of progress Work done with normal access to job, no impediments Work performed at normal working temperatures

Distribution of Time, %	Study Plans	Order Material	Receive and Store Material	Move to Point of Use	Tool Up	Measure	Install	Nonproductive	Total
	2	2	2	2	18	12	52	10	100

Labor Unit Table 25.

Fittings, Conduit, Rigid Steel				
Size, in	Man-hours per each			
	Expansion fittings	Nipples	Couplings	Clamps
1/2	0.37	0.11	0.04	0.03
3/4	0.39	0.12	0.05	0.04
1	0.40	0.14	0.07	0.05
1-1/4	0.42	0.15	0.09	0.07
1-1/2	0.44	0.17	0.10	0.09
2	0.48	0.20	0.13	0.12
2-1/2	0.58	0.31	0.24	0.23
3	0.63	0.36	0.29	0.28
3-1/2	0.68	0.41	0.34	0.33
4	0.73	0.46	0.39	0.38
5	0.85	0.55	0.45	0.43

Factoring

Conditions of Usage

Nipples include straight, chase, offset types
Pipe dope application included where needed

Conditions of Installation

Work done by proper number of average journeymen using proper tools
Work installed in a local area, at ground level, and at a height of not over 14 ft
Work done with normal concentrations of material, with normal material delivery, and at normal rate of progress
Work done with normal access to job, no impediments
Work performed at normal working temperatures

Distribution of Time, %

Study Plans	Order Material	Receive and Store Material	Move to Point of Use	Tool Up	Measure	Install	Nonproductive	Total
3	3	5	5	15	5	54	10	100

Labor Unit Table 26.

Fixtures, Lighting

Size, ft	Description	Type of construction	Man-hours per each
4	Fixture, fluorescent commercial 2/48 (R)	Free air	1.10
4	Fixture, fluorescent commercial 4/48 (R)	Free air	1.35
8	Fixture, fluorescent commercial 2/96 (S)	Free air	1.65
8	Fixture, fluorescent commercial 4/96 (S)	Free air	1.95
4	Fixture, fluorescent industrial 2/48 (R)	Free air	0.80
4	Fixture, fluorescent industrial 4/48 (R)	Free air	1.00
8	Fixture, fluorescent industrial 2/96 (S)	Free air	1.30
8	Fixture, fluorescent industrial 4/96 (S)	Free air	1.60
4	Fixture, fluorescent layin 4/48 (R)	Recessed	1.25
8	Fixture, fluorescent layin 4/96 (S)	Recessed	1.90
4	Fixture, fluorescent strip 2/48 (R)	Free air	0.80
8	Fixture, fluorescent strip 2/96 (S)	Free air	1.30
	Fixture, incandescent, recessed 1/150 Watt	Recessed	0.90
	Fixture, incandescent, surface 2/100 Watt	Free air	0.70
	Fixture, incandescent, 1/PC 1/300 Watt	Free air	0.60
	Fixture, MV, industrial 1/250 Watt	Free air	1.20
	Fixture, MV, industrial 1/400 Watt	Free air	1.50

Factoring

Conditions of Usage

Fastenings not included
Units include lamping and connecting

Conditions of Installation

Work done by proper number of average journeymen using proper tools
Work installed in a local area, at ground level, and at a height of not over 14 ft
Work done with normal concentrations of material, with normal material delivery, and at normal rate of progress
Work done with normal access to job, no impediments
Work performed at normal working temperatures

Distribution of Time, %

Study Plans	Order Material	Receive and Store Material	Move to Point of Use	Tool Up	Measure	Install	Nonproductive	Total
5	2	5	5	5	5	63	10	100

Labor Unit Table 27.

Knockout, Cutting			
Size, in	Man-hours per each		
	Hand	Holesaw	Hydraulic
1/2	0.20	0.15	0.18
3/4	0.25	0.17	0.21
1	0.27	0.20	0.21
1-1/4	0.29	0.22	0.21
1-1/2	0.31	0.25	0.21
2	0.40	0.30	0.25
2-1/2	0.47	0.40	0.30
3	0.55	0.50	0.35
3-1/2	0.65	0.60	0.40
4	0.75	0.70	0.45

Factoring	
Conditions of Usage	Units based on using electric drill for 1/2-in pilot hole drilled for hand and hydraulic operations
Conditions of Installation	Work done by proper number of average journeymen using proper tools Work installed in a local area, at ground level, and at a height of not over 14 ft Work done with normal concentrations of material, with normal material delivery, and at normal rate of progress Work done with normal access to job, no impediments Work performed at normal working temperatures

Distribution of Time, %	Study Plans	Order Material	Receive and Store Material	Move to Point of Use	Tool Up	Measure	Install	Nonproductive	Total
	2	2	2	2	20	10	62	10	100

Labor Unit Table 28.

Lugs, Solderless, Bolted, Compression	
Size no	Man-hours per each
8	0.15
6	0.18
4	0.19
3	0.19
2	0.20
1	0.22
1/0	0.23
2/0	0.25
3/0	0.27
4/0	0.30
250	0.37
300	0.41
350	0.46
400	0.52
500	0.65

Factoring

Conditions of Usage: Cutting, stripping, and connecting included

Conditions of Installation:

Work done by proper number of average journeymen using proper tools
Work installed in a local area, at ground level, and at a height of not over 14 ft
Work done with normal concentrations of material, with normal material delivery, and at normal rate of progress
Work done with normal access to job, no impediments
Work performed at normal working temperatures

Distribution of Time, %	Study Plans	Order Material	Receive and Store Material	Move to Point of Use	Tool Up	Measure	Install	Nonproductive	Total
	2	2	2	2	6	6	70	10	100

Labor Unit Table 29.

Panelboard, Circuit Breaker Bolted Type

Size of branch	Neutral bar time	Man-hours per each SED frame	SFH frame	SGH frame	SGL6 frame	SKH8 frame	SKH12 frame
70	0.14	2.20	2.50				
90	0.16	2.30	2.60				
100	0.18	2.30	2.60				
125	0.21	2.45	2.75	3.00	3.70	5.15	
150	0.27	2.60	2.90	3.15	3.85	5.25	
175	0.33	2.75	3.05	3.30	4.00	5.40	
200	0.38	2.90	3.20	3.45	4.15	5.75	
225	0.46	3.30	3.60	3.75	4.45	5.90	
250	0.53	3.40	3.70	3.95	4.65	6.10	
300	0.62	4.15	4.45	4.70	5.40	6.35	
325	0.71	4.60	4.95	5.20	5.90	6.60	
350	0.79	5.20	5.50	5.75	6.50	6.90	
400	0.97	5.40	5.70	5.95	7.00	8.00	9.10
500	1.50				8.50	9.60	10.70
600	2.00				10.20	11.10	12.40
800						12.00	14.50
Cabinet time....		1.03	1.35	1.50	2.35		

Note: Breaker voltage-480V
Time shown in man-hours.
All breakers molded case.

Note: SED breakers, 15-150A, 14K-100K IC
SFH breakers, 70-250A, 25K-100K IC
SGH breakers, 125-400A, 35K-100K IC
The total panelboard labor is a sum of the man-hours for all branches, labor for main lugs included.
SGL6 breakers, 250-600A, 65K-100K IC
SKH8 breakers, 300-800A, 50K-100K IC
SKH12 breakers, 600-1200A, 50K-100K IC

Factoring

Conditions of Usage

Labor for extra supports not included
Units apply to flush or surface installation
Labor for wire terminals included
Fastening labor not included. Conduit terminals not included.
Directory time included. Trim time included.

Conditions of Installation

Work done by proper number of average journeymen using proper tools
Work installed in a local area, at ground level, and at a height of not over 14 ft
Work done with normal concentrations of material, with normal material delivery, and at normal rate of progress
Work done with normal access to job, no impediments
Work performed at normal working temperatures

Distribution of Time, %

Study Plans	Order Material	Receive and Store Material	Move to Point of Use	Tool Up	Measure	Install	Nonproductive	Total
8	2	4	4	3	3	66	10	100

Labor Unit Table 30.

Panelboard, Lighting

No. of circuits	Cabinet trim time	Neutral bar time	Terminal time	Total man-hours each
4	2.55	0.22	1.28	4.00
6	2.75	0.25	1.92	5.00
8	2.75	0.28	2.56	5.60
10	2.75	0.35	3.20	6.30
12	2.75	0.40	3.84	7.00
14	2.95	0.50	4.48	8.00
16	2.95	0.55	5.12	8.60
18	3.15	0.60	5.76	9.50
20	3.15	0.65	6.40	10.20
22	3.15	0.70	7.04	11.00
24	3.15	0.80	7.68	11.70
26	3.45	0.90	8.32	12.70
28	3.95	1.10	8.96	14.00
30	4.15	1.25	9.60	15.00
32	4.15	1.35	10.24	15.80
34	4.15	1.50	10.88	16.50
36	4.15	1.60	11.52	17.30
38	4.35	1.70	12.16	18.20
40	4.35	1.80	12.80	19.00
42	4.55	2.00	13.44	20.00

Note: Breaker voltage-240V, 10,000 AIC
Time shown in man-hours.

Factoring

Conditions of Usage

Labor for extra supports not included
Units apply to flush or surface installation
Labor for wire terminals included
Fastening labor not included. Conduit terminals not included.
Directory time included. Trim time included.

Conditions of Installation

Work done by proper number of average journeymen using proper tools
Work installed in a local area, at ground level, and at a height of not over 14 ft
Work done with normal concentrations of material, with normal material delivery, and at normal rate of progress
Work done with normal access to job, no impediments
Work performed at normal working temperatures

Distribution of Time, %

Study Plans	Order Material	Receive and Store Material	Move to Point of Use	Tool Up	Measure	Install	Nonproductive	Total
8	2	4	4	3	3	66	10	100

Labor Unit Table 31.

Panelboard, Residential

No. of circuits	Cabinet trim time	Neutral bar time	Terminal time	Total man-hours each
2	1.50	0.20	0.64	2.30
4	1.75	0.22	1.28	3.25
6	2.15	0.25	1.92	4.30
8	2.15	0.28	2.56	5.00
10	2.15	0.35	3.20	5.70
12	2.15	0.40	3.84	6.40
14	2.35	0.50	4.48	7.30
16	2.35	0.55	5.12	8.00
18	2.35	0.60	5.76	8.70
20	2.55	0.65	6.40	9.60
22	2.55	0.70	7.04	10.30
24	2.55	0.80	7.68	11.00
26	2.75	0.90	8.32	12.00
28	2.75	1.10	8.96	12.80
30	2.75	1.25	9.60	13.60
32	3.45	1.35	10.24	15.00
34	3.45	1.50	10.88	15.80
36	3.45	1.60	11.52	16.60
38	3.75	1.70	12.16	17.60
40	3.75	1.80	12.80	18.40
42	3.75	2.00	13.44	19.20

Note: Time shown in man-hours.

Factoring

Conditions of Usage

Labor for extra supports not included
Units apply to flush or surface installation
Labor for wire terminals included
Fastening labor not included. Conduit terminals not included.
Directory time included. Trim time included.

Conditions of Installation

Work done by proper number of average journeymen using proper tools
Work installed in a local area. at ground level, and at a height of not over 14 ft
Work done with normal concentrations of material. with normal material delivery. and at normal rate of progress
Work done with normal access to job. no impediments
Work performed at normal working temperatures

Distribution of Time, %

Study Plans	Order Material	Receive and Store Material	Move to Point of Use	Tool Up	Measure	Install	Nonproductive	Total
8	2	4	4	3	3	66	10	100

Labor Unit Table 32.

Sleeves			
Size, in	Description	Type of construction	Man-hours per each
1/2-2	Sleeve,conduit/flange	Deck	1.10
2-1/2-4	Sleeve,conduit/flange	Deck	1.40
1/2-2	Sleeve,sheet metal	Deck	0.60
2-1/2-4	Sleeve,sheet metal	Deck	0.90
4x10	Sleeve,box opening	Wall	2.00

Factoring	
Conditions of Usage	Unit based on attaching to wood forms
Conditions of Installation	Work done by proper number of average journeymen using proper tools Work installed in a local area, at ground level, and at a height of not over 14 ft Work done with normal concentrations of material, with normal material delivery, and at normal rate of progress Work done with normal access to job, no impediments Work performed at normal working temperatures

Distribution of Time, %	Study Plans	Order Material	Receive and Store Material	Move to Point of Use	Tool Up	Measure	Install	Nonproductive	Total
	2	2	2	2	4	10	68	10	100

Labor Unit Table 33.

Starter, Motor, Magnetic, 250V, 3-Pole		
Size, hp	Starter Size	Man-hours per each
1.0	00	2.20
1.5	00	2.20
2.0	0	2.30
3.0	0	2.30
5.0	1	2.45
7.5	1	2.70
10	2	3.05
15	2	3.45
20	3	4.45
25	3	4.70
30	3	5.10
40	4	7.70
50	4	8.10
60	4	13.45
75	4	14.50
100	5	15.50

Factoring

Conditions of Usage

Checking motor rotation included
Handling, placing and connecting line, load,and control included
Conduit terminals not included. Fastening not included
Two-man installation 60 hp and above
Units based on general purpose enclosure

Conditions of Installation

Work done by proper number of average journeymen using proper tools
Work installed in a local area, at ground level, and at a height of not over 14 ft
Work done with normal concentrations of material, with normal material delivery, and at normal rate of progress
Work done with normal access to job, no impediments
Work performed at normal working temperatures

Distribution of Time, %	Study Plans	Order Material	Receive and Store Material	Move to Point of Use	Tool Up	Measure	Install	Nonproductive	Total
	4	2	3	5	5	5	66	10	100

Labor Unit Table 34.

Started, Motor, Manual		
Size, hp	Description	Man-hours per each
1	Starter, fractional horsepower	1.50
2	Starter, nema 1 enclosure 1 PH	2.00
3	Starter, nema 1 enclosure 1 PH	2.00
5	Starter, nema 1 enclosure 1 PH	2.00
1	Starter, nema 1 enclosure 3 PH	2.50
2	Starter, nema 1 enclosure 3 PH	2.50

Factoring

Conditions of Usage
Checking motor rotation included
Handling, placing and connecting included
Conduit terminals not included
Fastening labor not included
Units based on general purpose enclosure

Conditions of Installation
Work done by proper number of average journeymen using proper tools
Work installed in a local area, at ground level, and at a height of not over 14 ft
Work done with normal concentrations of material, with normal material delivery, and at normal rate of progress
Work done with normal access to job, no impediments
Work performed at normal working temperatures

Distribution of Time, %	Study Plans	Order Material	Receive and Store Material	Move to Point of Use	Tool Up	Measure	Install	Nonproductive	Total
	4	2	3	5	5	5	66	10	100

Labor Unit Table 35.

Switchboard, Air Circuit Breaker Type			
Size, A	Description	Man-hours per each	
		NB frame	P B frame
700	Breaker, circuit, 3P, 600V, 75,000AIC	18.00	23.00
800	Breaker, circuit, 3P, 600V, 75,000AIC	20.20	27.50
1000	Breaker, circuit, 3P, 600V, 75,000AIC	24.10	32.50
1200	Breaker, circuit, 3P, 600V, 75,000AIC	27.00	37.30
1600	Breaker, circuit, 3P, 600V, 75,000AIC		42.80
2000	Breaker, circuit, 3P, 600V, 75,000AIC		47.50
2500	Breaker, circuit, 3P, 600V, 75,000AIC		53.00
3000	Breaker, circuit, 3P, 600V, 75,000AIC		58.00
	Breakers, drawout type (cabinet time)	19.75	19.75

Note: The total panel board labor is a sum of the man-hours for all branches, labor for main lugs included

Conditions of Usage

Labor for extra supports not included
Units apply to freestanding cubicle
Labor for wire terminals included
Conduit terminals not included

Conditions of Installations

Work done by proper number of average journeymen using proper tools
Work installed in a local area, at ground level, and at a height of not over 14 ft
Work done with normal concentrations of material, with normal material delivery, and at normal rate of progress
Work done with normal access to job, no impediments
Work performed at normal working temperatures

Distribution of Time, %

Study Plans	Order Material	Receive and Store Material	Move to Point of Use	Tool Up	Measure	Install	Nonproductive	Total
8	2	4	4	8	4	60	10	100

Labor Unit Table 36.

Switches, Safety, 240V, NEMA 1

Size, A	Man-hours per each		
	2-Pole	3-Pole	4-Pole
30	1.25	1.60	1.70
60	1.85	2.00	2.25
100	2.70	3.10	3.60
200	4.80	5.40	6.00
400	9.00	10.40	11.10
600	11.00	13.00	15.00

Factoring

Conditions of Usage

- Fastening labor not included
- Conduit terminals not included
- Two-man installation for 400-A and 600-A switches
- Handling, placing, and connecting included
- Units based on general duty type switches

Conditions of Installations

- Work done by proper number of average journeymen using proper tools
- Work installed in a local area, at ground level, and at a height of not over 14 ft
- Work done with normal concentrations of material, with normal material delivery, and at normal rate of progress
- Work done with normal access to job, no impediments
- Work performed at normal working temperatures

Distribution of Time, %

Study Plans	Order Material	Receive and Store Material	Move to Point of Use	Tool Up	Measure	Install	Nonproductive	Total
4	2	3	3	5	5	68	10	100

Labor Unit Table 37.

Terminals, Conduit, Rigid Steel

Size, in	Man-hours per each: Hub terminal construction	Man-hours per each: Free-air construction
1/2	0.13	0.17
3/4	0.17	0.21
1	0.22	0.28
1-1/4	0.29	0.36
1-1/2	0.34	0.43
2	0.44	0.56
2-1/2	0.67	0.86
3	0.85	1.10
3-1/2	1.02	1.31
4	1.19	1.55
5	1.60	2.10

Factoring

Reduce units for nippling up 17%
Factor for different types of construction
(see Form E-19)

Conditions of Usage

Units predicated on use of mechanical threaders
Offsets and bends not included in units
Hub terminal does not include locknuts or bushings
Terminals include one bushing and two locknuts
Terminals predicated on 3/4 of one cut and thread

Conditions of Installation

Work done by proper number of average journeymen using proper tools
Work installed in a local area, at ground level, and at a height of not over 14 ft
Work done with normal concentrations of material, with normal material delivery, and at normal rate of progress
Work done with normal access to job, no impediments
Work performed at normal working temperatures

Distribution of Time, %

Study Plans	Order Material	Receive and Store Material	Move to Point of Use	Tool Up	Measure	Install	Nonproductive	Total
2	2	2	2	18	12	52	10	100

Labor Unit Table 38.

Wire, Aluminum, THHN, 600V Drawn in Conduit

Size	Man-hours per ft 2-wire pull	3-wire pull	4-wire pull
3 AWG	0.0156	0.0117	0.0098
2 AWG	0.0170	0.0128	0.0110
1 AWG	0.0192	0.0144	0.0126
1/0 AWG	0.0247	0.0185	0.0165
2/0 AWG	0.0282	0.0212	0.0188
3/0 AWG	0.0330	0.0248	0.0218
4/0 AWG	0.0385	0.0289	0.0251
250 MCM	0.0440	0.0330	0.0287
300 MCM	0.0480	0.0360	0.0315
350 MCM	0.0520	0.0390	0.0339
400 MCM	0.0580	0.0435	0.0383
500 MCM	0.0630	0.0473	0.0414
600 MCM	0.0775	0.0581	0.0524
750 MCM	0.0850	0.0638	0.0563

Note: Units are based on labor for 1 foot of 1 wire in a 4-wire pull of 100 ft in length, for example, 100 ft of (4 No. 2) = 100 x 0110 x 4 = 4.40 man-hours

Factoring

Factors for pulls of other lengths: 50 ft — 1.50 Multiplier
150 ft — 0.85 Multiplier
200 ft — 0.70 Multiplier
300 ft — 0.60 Multiplier

Factor for parallel pulls: 2 pulls — 0.90 multiplier; 3 pulls — 0.85; 4 pulls — 0.80

Conditions of Usage

Units based on using mechanical pulling equipment
Wire is field measured, but cut to length off job
Multiple wire runs delivered to job on reels, cut in parallel lengths
Terminations not included in units

Conditions of Installation

Work done by proper number of average journeymen using proper tools
Work installed in a local area, at ground level, and at a height of not over 14 ft
Work done with normal concentrations of material, with normal material delivery, and at normal rate of progress
Work done with normal access to job, no impediments
Work performed at normal working temperatures

Distribution of Time, %

Study Plans	Order Material	Receive and Store Material	Move to Point of Use	Tool Up	Measure	Install	Nonproductive	Total
2	2	5	5	28	3	45	10	100

Labor Unit Table 39.

Wire, Copper, THHN Drawn in Conduit			
Size	Man-hours per ft		
	2-wire pull	3-wire pull	4-wire pull
14 AWG	0.0066	0.0050	0.0045
12 AWG	0.0077	0.0059	0.0045
10 AWG	0.0095	0.0072	0.0063
8 AWG	0.0132	0.0099	0.0081
6 AWG	0.0150	0.0113	0.0090
4 AWG	0.0169	0.0134	0.0112
2 AWG	0.0203	0.0162	0.0139
1 AWG	0.0230	0.0182	0.0160
1/0 AWG	0.0297	0.0235	0.0209
2/0 AWG	0.0338	0.0268	0.0237
3/0 AWG	0.0396	0.0314	0.0276
4/0 AWG	0.0462	0.0366	0.0318
250 MCM	0.0528	0.0418	0.0363
350 MCM	0.0625	0.0494	0.0430
500 MCM	0.0756	0.0600	0.0525

Note: Units are based on labor for 1 foot of 1 wire in a 4-wire pull of 100 ft in length, for example, 100 ft of (4 No. 2) = 100 x 0.0139x4 = 5.56 man-hours

Factoring

Factors for pulls of other lengths: 50 ft — 1.50 Multiplier
150 ft — 0.85 Multiplier
200 ft — 0.70 Multiplier
300 ft — 0.60 Multiplier
Factor for parallel pulls: 2 pulls — 0.90 multiplier; 3 pulls — 0.85; 4 pulls — 0.80

Conditions of Usage

Units based on using mechanical pulling equipment
Wire is field measured, but cut to length off job
Multiple wire runs delivered to job on reels, cut in parallel lengths
Terminations not included in units

Conditions of Installation

Work done by proper number of average journeymen using proper tools
Work installed in a local area, at ground level, and at a height of not over 14 ft
Work done with normal concentrations of material, with normal material delivery, and at normal rate of progress
Work done with normal access to job, no impediments
Work performed at normal working temperatures

Distribution of Time, %

Study Plans	Order Material	Receive and Store Material	Move to Point of Use	Tool Up	Measure	Install	Nonproductive	Total
2	2	5	5	28	3	45	10	100

Labor Unit Table 40.

Index

ABOUT THE AUTHORS

RALPH E. JOHNSON, a graduate of Colorado School Mines, is a pioneer in the development of computerized estimating. He is a member of the National Electrical Contractors Association (NECA) and Academy of Electrical Contractors. He is the author of *Electrical Construction Cost Manual* and *Electrical Contracting Business Handbook*, published by McGraw-Hill. He is retired president of Sturgeon Electric Co. in Denver, Colorado.

GENE WHITSON is a CPA member of the International Accountants Society, a leader in the computerization of the electrical contracting business and a retired vice president of Cannon-Wendt Electric Company in Phoenix, Arizona.